Guided Inquiry
for General Chemistry

FIRST EDITION

J. Hugh Broome, Ph.D. and Mary A. Mackey, Ph.D.

cognella®

SAN DIEGO

Bassim Hamadeh, CEO and Publisher
Angela Schultz, Senior Field Acquisitions Editor
Michelle Piehl, Senior Project Editor
Celeste Paed, Associate Production Editor
Jess Estrella, Senior Graphic Designer
Trey Soto, Licensing Coordinator
Kim Scott/Bumpy Design, Interior Designer
Natalie Piccotti, Director of Marketing
Kassie Graves, Vice President of Editorial
Jamie Giganti, Director of Academic Publishing

Printed in the United States of America.

cognella® | ACADEMIC PUBLISHING

3970 Sorrento Valley Blvd., Ste. 500, San Diego, CA 92121

Brief Contents

Detailed Contents

CHAPTER 1

Math for Chemists

How do I convert between units of the same measurement?

How do I convert from one type of measurement to another?

How many decimal places should I keep once I have an answer?

Learning Outcomes

- Perform calculations and conversions using the metric system.
- Report results in scientific notation.
- Convert between different types of units (e.g., mass to volume).
- Apply the concept of significant figures to calculations.
- Apply algebra and arithmetic to chemistry application problems.

Prerequisite Knowledge

- None!

Background Information and Questions

Mathematics is an important aspect of chemistry. In fact, chemistry is basically applied mathematics. Chemistry is difficult to learn because it is not a stand-alone subject; it builds chemical concepts onto mathematical ones. This lesson is designed to arm you with the mathematical skills necessary to succeed in chemistry. In chemistry, scientists make many measurements; therefore, this lesson begins with units. The metric system is an excellent place to get started; it is the most common measurement system in science. The metric system utilizes the International System of Units (SI units) as a basis for measurement (**Table 1.1**).

TABLE 1.1 SI Units

Name of SI Unit	Symbol	Physical Quantity
meter	m	Length
kilogram	kg	Mass
second	s	Time
kelvin	K	Temperature
ampere	A	Electric current
mole	mol	Amount of substance
candela	cd	Luminous intensity

Other units can be derived from these seven units as needed, such as the hertz (1/s) to measure frequency or the watt ($kg \cdot m^2 \cdot s^{-3}$) to measure power. Metric system prefixes, such as milli- or kilo-, are used to report numbers in fewer digits when required for measurements. One should pick the appropriate prefix for the measurement being made. For example, it would be silly to report driving distances in meters instead of kilometers. These prefixes are based on a decimal system, and the most common ones used in chemistry are shown in **Table 1.2**.

TABLE 1.2 Most Common Metric System Prefixes

Prefix	Symbol	Power	Factor
giga	G	10^9	1,000,000,000
mega	M	10^6	1,000,000
kilo	k	10^3	1,000
—	—	1	1
centi	c	10^{-2}	0.01
milli	m	10^{-3}	0.001
micro	μ	10^{-6}	0.000001
nano	n	10^{-9}	0.000000001

Examples (parts directly from table in bold):

$$1 \textbf{ kilo}\text{meter} = 1 \textbf{ km} = 1 \times \mathbf{10^3} \text{ m} = \mathbf{1,000} \text{ meters}$$

To go in reverse, one must look at the inverse of the numbers.

$$1 \text{ meter} = 1 \text{ m} = 1 \times 10^{-3} \text{ km} = 0.001 \text{ kilometers}$$

Another example is that 1 centimeter (cm) is equal to 1×10^{-2} meters, which is the same as 0.01 meters. Conversely, 1 meter is equal to 1×10^2 centimeters, which is the same as 100 centimeters. This relationship—that of one thing being equivalent to another—can be used to convert from one unit to another in a process called *dimensional analysis*.

For example, how many kilograms are in 3,500 grams? In the following example, notice that grams (g) appears in both a numerator and a denominator, so you are essentially dividing grams by grams to get the number 1, allowing grams to "cancel out" and leave only kilograms (kg) as a unit for our answer.

$$\frac{3,500 \text{ g}}{1} \times \frac{1 \text{ kg}}{1 \times 10^3 \text{ g}} = 3.5 \text{ kg}$$

Here is another dimensional analysis example illustrating two different methods. How many centimeters are in 2.56 kilometers? There is more than one way to calculate this—take note of the two methods below. Both use an equivalent number (1 km is equal to 1×10^3 meters, just like 1×10^2 cm is equal to 1 meter) in both the numerator and denominator to obtain the desired unit.

$$\text{Method 1:} \frac{25.6 \text{ km}}{1} \times \frac{1 \times 10^3 \text{ m}}{1 \text{ km}} \times \frac{1 \times 10^2 \text{ cm}}{1 \text{ m}} = 2,560,000 \text{ cm}$$

$$\frac{25.6 \cancel{\text{ km}}}{1} \times \frac{1 \times 10^3 \cancel{\text{ m}}}{1 \cancel{\text{ km}}} \times \frac{1 \times 10^2 \cancel{\text{ cm}}}{1 \cancel{\text{ m}}} = 2,560,000 \cancel{\text{ cm}}$$

$$\text{Method 2:} \frac{25.6 \text{ km}}{1} \times \frac{\text{m}}{1 \times 10^{-3} \text{ km}} \times \frac{1 \text{ cm}}{1 \times 10^{-2} \text{ m}} = 2,560,000 \text{ cm}$$

$$\frac{25.6 \cancel{\text{ km}}}{1} \times \frac{\cancel{\text{m}}}{1 \times 10^{-3} \cancel{\text{ km}}} \times \frac{1 \cancel{\text{ cm}}}{1 \times 10^{-2} \cancel{\text{ m}}} = 2,560,000 \cancel{\text{ cm}}$$

A problem with reporting in a specific unit (centimeters in this example) is that the numbers can get too large or too small to record or understand at a glance. The number 2,560,000 would therefore be expressed in scientific notation, as follows:

$$2,560,000 \text{ cm} \rightarrow 2.56 \times 10^6 \text{ cm}$$

This result in scientific notation could be expanded as follows to show that the exponent denotes how many times the number must be multiplied by 10:

$$2.56 \times 10^6 \text{ cm} = 2.56 \times 10 \times 10 \times 10 \times 10 \times 10 \times 10 = 2{,}560{,}000 \text{ cm}$$

We can also get this number by moving the decimal place from its current location either left or right so that there is a single digit in the ones place. In this case we moved the decimal place six times to the left and used a positive six for our exponent to indicate that.

$$2560000 \text{ cm} \rightarrow 2.56 \times 10^6 \text{ cm}$$

If we moved the decimal place to the right, we would use a negative exponent, as we would be starting with a number less than 1. An example of the latter is shown below:

$$0.000055 \text{ g} \rightarrow 5.5 \times 10^{-5} \text{ g}$$

1. How many milligrams are in 0.070 g?

2. How many kilograms are in 0.070 g?

3. Express 0.0035 m in scientific notation.

4. Express 35,000 m in scientific notation.

Dimensional analysis is a useful tool for converting between many types of units. Regardless of the units, the same rules apply: ensure that the denominator and the numerator are equivalent, cross out units that appear on both the "top" and "bottom," and report your answer with the remaining units you have not crossed out.

■ For example, how many seconds are in a day?

$$\frac{1\text{ day}}{1}\times\frac{24\text{ hours}}{1\text{ day}}\times\frac{60\text{ minutes}}{1\text{ hour}}\times\frac{60\text{ seconds}}{1\text{ minute}}=86,400\text{ seconds}$$

■ How many meters are in 1 mile? Assume the only conversion between the imperial and metric systems you know is that 1 inch is equal to 2.54 cm.

$$\frac{1\text{ mile}}{1}\times\frac{5,280\text{ ft}}{1\text{ mile}}\times\frac{12\text{ in}}{1\text{ ft}}\times\frac{2.54\text{ cm}}{1\text{ in}}\times\frac{1\text{ m}}{100\text{ cm}}=1,609\text{ meters}$$

■ How fast is 5.00 km per hour in meters per second?

$$\frac{5.00\text{ km}}{\text{hr}}\times\frac{1000\text{ m}}{1\text{ km}}\times\frac{1\text{ hr}}{60\text{ min}}\times\frac{1\text{ min}}{60\text{ s}}=1.389\text{ m}/\text{s}$$

5. How much does a 20.0-pound exercise ball weigh in kilograms, given that 1 kg is equal to 2.2 pounds (lbs)?

All this arithmetic brings up the question of how many decimal places to use in our answer. Why did we report 1.389 m/s instead of 1.39 or 1.3889 in the previous example? The answer is significant figures.

Significant figures are the numerals that are used to indicate the uncertainty associated with a measurement. For example, if you are using a ruler to measure something that lands between 20 cm and 21 cm, you might estimate that it is 20.5 cm. That last digit, the 5, is uncertain because you estimated. It's good practice to always (and only) estimate one decimal place past the listed values. Lab equipment will do this for you, so if a scale in the lab reads 42.60 g, you can assume that the actual weight is between 42.59 g and 42.61 g—the last digit is uncertain. When working with significant figures, it is important not to add additional uncertainty to the reported value. That concept is where the basic rules of significant figures apply:

■ Nonzero digits are significant.
  ■ 427.5 has four significant figures.
■ Ignore leading zeros.
  ■ 0.000567 has three significant figures; the four zeros at the front of the number are "leading" zeros.

▪ Ignore trailing zeros unless they follow a decimal point, which indicates they were accurately measured.
  ▪ 2,400 has two significant figures, whereas 24.00 has four significant figures.
▪ Zeros between significant figures (captive zeros) are also significant.
  ▪ 906.05 has five significant figures.
▪ *And a trick!* The number 300 would, at first glance, only have one significant figure (the 3). However, it is possible for 300 to have three significant figures if it is an "exact number" that has been counted. For example, if I have 300 students, then I have *exactly* 300 students—there is no uncertainty.

6. How many significant figures are in the following numbers?

   7,521   1.000   0.62 9.9×10^4 5,004.78

Different measurements will produce different numbers of significant figures. In this book we will do many calculations with these measurements. Let's talk about the rules regarding arithmetic and "sig figs":

▪ For *addition and subtraction*, the answer should have the same number of decimal places as the least accurate measurement (number with least decimal places).
  ▪ 62.0 + 0.775 = 62.8
▪ For *multiplication and division*, the answer should have the same number of significant figures as the least accurate measurement (number with least significant figures).
  ▪ 23.185 × 1.41 = 32.7
▪ For *logarithms and antilogarithms*, the answer should have the same number of decimal points to the right as there were significant figures in the original number. In chemistry we will assume that all logs are base 10 unless otherwise noted.
  ▪ Log (6.500) = 0.8129

7. Answer the following using significant figures:

  ▪ $55.3 \div 7.004 =$

  ▪ $300.0 - 1.475 =$

  ▪ $Log(3.3) =$

$$\frac{54.1 + 17.50}{6.331} =$$

Math skills are important in chemistry courses, and the concept and implementation of the logarithm is worth review. The logarithm is the inverse of exponentiation in the sense that division is the inverse of multiplication. An exponent indicates how many times to the number in multiplication (10^4 = 10 × 10 × 10 × 10 = 10,000). Therefore, a logarithm indicates which exponent produced the number in question ($10^?$ = 10,000). For example:

$$10,000 = 10 \times 10 \times 10 \times 10 = 10^? \text{ ... therefore ... } log_{10}\left(10,000\right) = 4$$

Lastly, a brief refresher on algebra and the order of operations is vital for success in chemistry. Algebra differs from arithmetic in that it often utilizes letters to represent unknown values. Often, the goal of basic algebra is to "solve for" that unknown and determine its numerical value. This is accomplished by manipulating both sides of the algebraic equation, keeping in mind the following order of operations: *p*arentheses, *e*xponents, *m*ultiplication, *d*ivision, *a*ddition, and *s*ubtraction. This is often abbreviated as PEMDAS in college algebra courses. The steps for a sample algebra problem are as follows:

Step 1) $5x + 3 = \dfrac{3^2}{(9-7)}$

Step 2) $5x + 3 = \dfrac{3^2}{(2)}$

Step 3) $5x + 3 = \dfrac{9}{(2)}$

Step 4) $2\left(5x+3\right) = 9$

Step 5) $10x + 6 = 9$

Step 6) $10x = 3$

Step 7) $x = \dfrac{3}{10}$

Step 8) $x = 0.3\overline{3}$

Review Questions

8. You are using a scale to measure sand, and the reported weight is 0.0005 kg. Rather than kilograms, which unit would be better to report this weight in, and why?

9. SI derived units are units that are derived using the seven base SI units. If a newton (N) and a joule (J) are defined as ...

$$N = \frac{m \times kg}{s^2} \qquad J = N \times m$$

... what would be the units for joules using only the seven base SI units?

Application Questions

Use significant figures and scientific notation (if the number of sig figs indicates you should) for ALL calculations!

10. How many nanometers are in 3.4 km?

11. How many kilograms are in 3.605×10^8 mg?

12. The speed of light is approximately 3.00×10^8 m/s. What is this in miles per hour?
    Hint! Recall that 1 inch = 2.54 cm.

13. One furlong per fortnight is a speed only just noticeable by the naked eye. If a furlong is defined as ⅛ of a mile, or 220 yards, and the fortnight is 14 days, what is this speed in m/s?
    Hint! Recall that 1 inch = 2.54 cm.

14. $\log(x) = 5.3$, therefore $x = ?$

15. $3x - 10 = \dfrac{25}{y}$, assume $y = 5$, then $x = ?$

16. $\dfrac{8x - 3z}{x^2} = 5 - y$, assume $y = 3$ and $z = 2$, then $x = ?$

CHAPTER 2

Matter

What is matter?

In what way do we classify matter?

How do pressure and temperature relate to the phases of matter?

Learning Outcomes

- Assign matter to one of its classifications.
- Identify chemical and physical properties of matter.
- Discuss how the phases of matter relate to kinetic energy.
- Utilize a phase diagram to determine how pressure and temperature affect the phases of matter.

Prerequisite Knowledge

- None!

Background Information and Questions

Matter is the term used to refer to any object that has both mass and volume. Mass is a measure of the amount of matter, whereas volume is a measure of space. Essentially, matter is anything that has mass and takes up space. Defining matter and discussing its classifications, properties, and phases is essentially a good introduction to the field of chemistry.

Matter can be classified into two main categories: pure substances and mixtures. Those two categories can be further divided; **Figure 2.1** illustrates how we classify matter. Pure substances are composed only of a single element or a single compound.

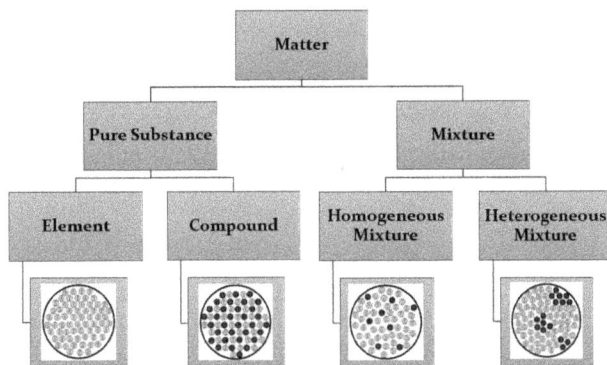

FIGURE 2.1 Classification of matter.

An element is the simplest form of matter; it cannot be broken via chemical reactions. A compound is two or more elements. Mixtures are composed of more than one element and/or compound. They can be heterogeneous or homogeneous. Homogeneous mixtures have an even dispersion of components, whereas heterogeneous mixtures have an uneven dispersion of mixture components.

1. How does a compound differ from a homogeneous mixture?

2. How would you classify the following? Choose from element, compound, homogenous mixture, or heterogenous mixture.

    (a) Gravel

    (b) Diamond

    (c) Water

    (d) Coffee

    (e) Milk

    (f) Copper

(g) Bronze

(h) Carbon dioxide

Scientists often describe matter based on its properties. There are two types of properties: physical and chemical. Chemical properties describe the "ability" of matter to undergo changes in the presence of certain other chemicals. Physical properties can be determined without changing the composition of the matter. Physical properties can be further divided into extensive physical properties and intensive physical properties. Extensive properties depend on the amount of matter, whereas intensive properties do not. **Figure 2.2** illustrates the relationships between physical, chemical, extensive, and intensive properties; it also provides some examples of each property. Chemical and physical changes occur as a result of these properties.

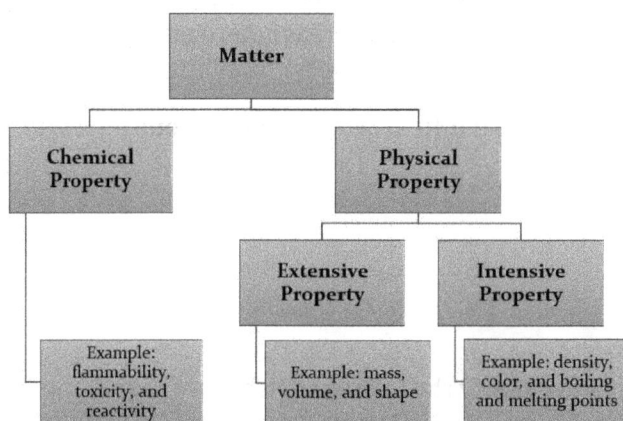

FIGURE 2.2 Physical, chemical, extensive, and intensive properties.

3. How would chemical and physical changes relate to chemical and physical properties, respectively?

Matter can exist in four different phases: solid, liquid, gas, or plasma. We will only focus on three phases: solid, liquid, and gas. As one moves from solid to liquid to gas, the kinetic energy of the molecules increases; gaseous atoms or molecules move much faster than atoms or molecules in a solid. Each phase of matter has unique properties as well.

Solids have a definite shape and volume. Liquids have a definite volume but an indefinite shape; they assume the shapes of their containers. Gases have both an indefinite shape and volume; they assume both the shape and volume of their containers. **Figure 2.3** illustrates the terms we use for describing the phase transitions in matter. As you move from a solid to a liquid, then to a gas, and finally to a plasma state, the kinetic energy of the atoms or molecules increases. Kinetic energy decreases as you move from a plasma to a solid.

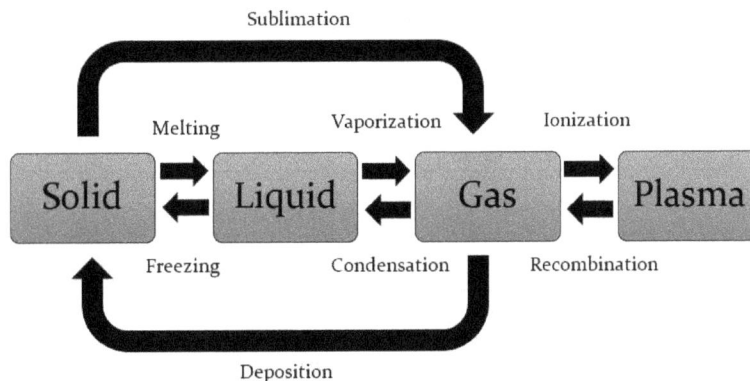

FIGURE 2.3 Phase transitions in matter.

4. Discuss how the physical properties of water change as it moves from a solid to a liquid.

Phase diagrams illustrate the temperatures and pressures at which certain phases exist for matter. **Figure 2.4** is a phase diagram that you can follow as you read about the components of a phase diagram. A phase diagram is plotted with temperature on the *x*-axis and pressure on the *y*-axis.

The solid phase lies on the left side of the graph at lower temperatures. The liquid phase lies to the right of the solid. The gas phase lies on the right side of the graph. The line in between the phases is known as the *phase boundary*; both phases exist in equilibrium at the phase boundary. The *triple point* is where all three phases exist in equilibrium. The *critical point* is the temperature and pressure where the density of the gas and liquid are equal. Above that temperature and pressure, a *supercritical fluid* exists. A supercritical fluid is indistinguishable as a gas or a liquid because the densities of a gas and a liquid are the same in a supercritical fluid.

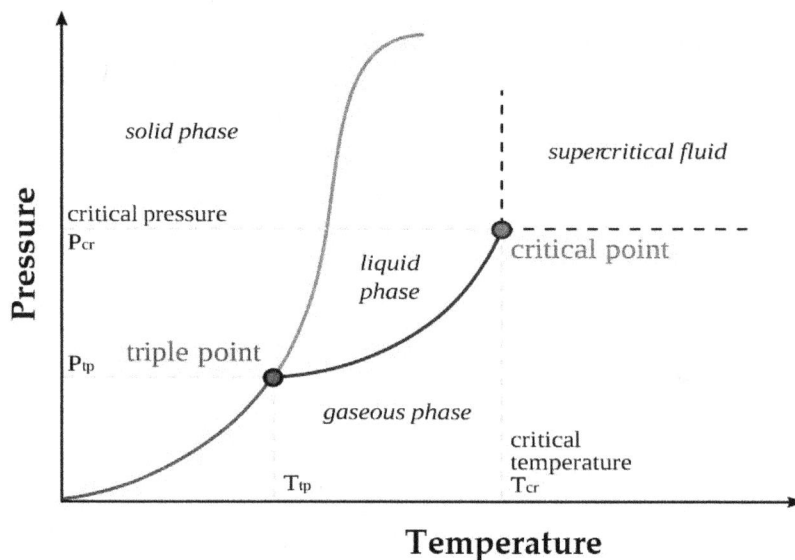

FIGURE 2.4 Phase diagram.

5. Describe the relationship of pressure and temperature to phases. In your description, make sure you include kinetic energy.

Review Questions

6. Which types of matter cannot be further simplified?

7. How do matter classifications compare/contrast to properties of matter?

Application Questions

8. Provide two examples of each of the following: elements, compounds, homogeneous mixtures, and heterogeneous mixtures.

9. Classify the following as a chemical or physical property.

   (a) Mass

   (b) Density

   (c) Electrical conductivity

   (d) Corrosion resistance

   (e) Hardness

10. Mass and volume are both extensive properties. Density, which is mass/volume, is an intensive property. Why?

11. Which of the following phase transitions requires an input of energy?

    (a) Sublimation

    (b) Condensation

(c)   Recombination

(d)   Freezing

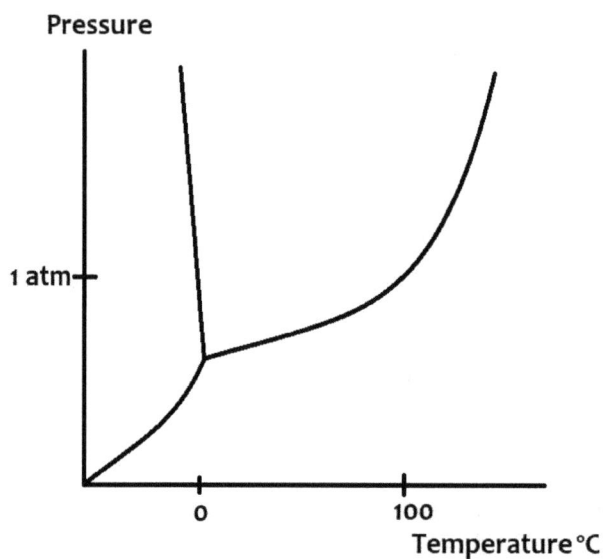

Pressure

1 atm

0   100

Temperature °C

Use the phase diagram of water above for the following questions.

12. What would be the phase of water at 0.5 atmosphere (atm) of pressure and 50 °C?

13. What phase or phases of water exist at 1 atm and 0 °C?

14. Estimate the conditions at which water would be at its triple point.

15. If you begin at roughly 0.1 atm and –50 Celsius, then warm your compound to 100 Celsius, which phase transition has your compound undergone?

Figure Credits

Fig. 2.1a: Copyright © by John Trombley (CC BY 4.0) at https://commons.wikimedia.org/wiki/File:Mixtures_and_Pure_Substances_2x2.svg.

Fig. 2.4: Copyright © by Maksim (CC BY-SA 3.0) at https://commons.wikimedia.org/wiki/File:Phase-diag.svg.

Fig. 2.5: Copyright © by Tornad (CC BY-SA 3.0) at https://commons.wikimedia.org/wiki/File:Diag_phase_eau.png.

CHAPTER 3

Atomic Structure

What makes up an atom?

What are isotopes, and why do they exist?

What types of elements form what types of ions?

Learning Outcomes

- Define the three subatomic particles.
- Identify the number of protons, neutrons, and electrons in an isotope.
- Utilize the average atomic mass formula to determine average atomic mass, percent abundance, or isotopic mass.
- Identify the charge of main group ions.

Prerequisite Knowledge

- None!

Background Information and Questions

The majority of an atom is empty space. Through a series of experiments, scientists have determined that atoms have a nucleus, which contains the majority of the atom's mass, and an electron cloud, which is the largest portion of the atom's volume. Hence, the portion of the atom with the largest volume contains the least amount of mass, meaning it is empty space. Atoms are composed of three subatomic particles: protons, neutrons, and electrons. Protons and neutrons are located in the nucleus; electrons are located in the electron cloud around the nucleus.

It is important to emphasize the scale of atoms and subatomic particles. Atoms are so small that their size is almost incomprehensible to us. If you expanded an atom to the size of your thumbnail, the nucleus would be the size of a cuticle cell.[1] The nucleus of an atom occupies a very small portion of an atom's volume but contains the majority of the atom's mass.

Atoms are identified as specific elements based on the number of protons; in other words, changing the number of protons changes the element. For example, all boron atoms have five protons, whereas carbon atoms have six protons. Even though all carbon atoms have six protons, some carbon atoms have six, seven, or eight neutrons. We can vary an atom's number of neutrons and electrons. This introduces isotopes and ions.

1. What three subatomic particles compose an atom? What are the particles' locations within the atom?

Isotopes are atoms of the same element; they have the same numbers of protons but different numbers of neutrons. Some of the more well-known isotopes are carbon-14, uranium-235, and hydrogen-2, which is known as deuterium. Certain isotopes can have unique properties; for example, some isotopes of an element can be radioactive. **Table 3.1** lists all of the carbon isotopes. The only naturally occurring isotopes of carbon are carbon-12, carbon-13, and carbon-14; all other carbon isotopes are synthetic, meaning that they are created in a laboratory setting. The right column of the table tells us the natural abundance of these isotopes. The table shows that carbon-12 has an abundance of 98.93%, which means that 98.93% of all carbon on Earth is carbon-12.

TABLE 3.1 Carbon Isotopes

Isotope	Number of Neutrons	Natural Abundance
Carbon-8	2	
Carbon-9	3	
Carbon-10	4	
Carbon-11	5	
Carbon-12	6	98.93%
Carbon-13	7	1.07%

(continued)

1 Marder, J. (December, 2010). Just Ask: What Would a Supersized Atom Look Like? PBS News Hour. https://www.pbs.org/newshour/science/just-ask-what-would-a-baseball-sized-atom-look-like

TABLE 3.1 Carbon Isotopes (*Continued*)

Isotope	Number of Neutrons	Natural Abundance
Carbon-14	8	Trace
Carbon-15	9	
Carbon-16	10	
Carbon-17	11	
Carbon-18	12	
Carbon-19	13	
Carbon-20	14	
Carbon-21	15	
Carbon-22	16	

Chemists use isotope notation to represent these isotopes. Before we discuss notation, let's talk about the terms *mass number* and *atomic number*. The mass number (A) is the number of protons and neutrons, whereas the atomic number (Z) is just the number of protons. If the atom is neutral, the atomic number can also represent the number of electrons, meaning it has an equal number of protons and electrons.

Isotope notation represents the various isotopes of an element; the following is a sample notation for the fictional element X. Next to the sample notation are the notations of the three naturally occurring carbon isotopes. Notice how the atomic number remains the same, while the mass number grows. This indicates an increasing number of neutrons, which are calculated by subtracting the atomic number from the mass number. The three naturally occurring isotopes have six, seven, and eight neutrons, from left to right.

$$_{Z}^{A}X, \quad _{6}^{12}C, \quad _{6}^{13}C, \quad _{6}^{14}C$$

2. Calculate the number of neutrons for the following isotopes.

(a) $_{7}^{14}N$

(b) $_{7}^{15}N$

(c) $_{92}^{238}U$

(d) $_{92}^{235}U$

The isotope notation is not what appears on the periodic table, so how do the notation and periodic table relate? **Figure 3.1** shows the block for the element boron. Arrows indicate where the components of that block appear in the notation. The atomic number of boron is 5, which appears in the denominator of the notation, and the symbol is B. Notice that the mass number does not appear in the periodic table. Also, notice that the "10.81" number appears on the periodic table but does not appear in the notation. We will discuss this number later in this lesson.

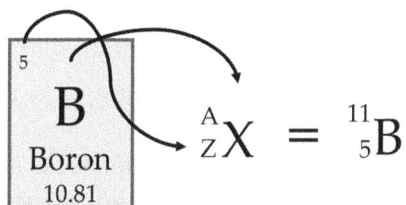

$$\,^{A}_{Z}X \;=\; \,^{11}_{5}B$$

5
B
Boron
10.81

FIGURE 3.1 Periodic block for boron.

One can also write this notation without including the atomic number, such as ^{11}B. Another way of notating isotopes is to simply write the full name followed by a hyphen and the mass number, such as boron-11. The atomic number is not needed in these notations because it is assumed you will look at the periodic table to determine the atomic number of the element or isotope.

3. Which element has 33 protons, 79 protons, and 108 protons?

Now, let's tackle the "10.81" number in the periodic block for boron in Figure 3.1. That number is known as the *atomic mass* or *average atomic mass.* It is calculated from all the naturally occurring isotopes using the following formula. This formula is a weighted average of the masses of all the naturally occurring isotopes of an element. Basically, the formula is the percent (%) abundance of each isotope as a decimal multiplied by the mass of that isotope. Then, you sum the products of the previous sentence's math operation.

Average atomic mass

$$= \left(\frac{\%\ \text{abundance of isotope 1}}{100} \times \text{mass of isotope 1} \right) + \left(\frac{\%\ \text{abundance of isotope 2}}{100} \times \text{mass of isotope 2} \right) + \dots$$

Let's try calculating the average atomic mass of magnesium. Magnesium has three naturally occurring isotopes: magnesium-24, magnesium-25, and magnesium-26. Magnesium-24 is 78.70% abundant and has an isotopic mass of 23.9850 atomic mass units (amu). Magnesium-25 has a natural abundance of 10.13% and an isotopic mass of 24.9858 amu. Finally, magnesium-26 is 11.17% abundant and has an isotopic mass of 25.9826 amu.

Average atomic mass

$$= \left(\frac{78.70}{100} \times 23.9850 \text{ amu} \right) + \left(\frac{10.13}{100} \times 24.9858 \text{ amu} \right) + \left(\frac{11.17}{100} \times 25.9826 \text{ amu} \right)$$

$$\text{Average atomic mass} = 24.3095 \approx 24.31 \text{ amu}$$

Notice that each set of parentheses in the calculation is for a naturally occurring isotope of magnesium. The answer, 24.31 amu, matches the average atomic mass on the periodic table. That is one way that you can check your calculations and work.

4. Calculate the average atomic mass of lithium, which has two isotopes: ^{6}Li and ^{7}Li.
    Lithium-6 has a mass of 6.0151 amu and an abundance of 7.59%; lithium-7 has an abundance of 92.41% and a mass of 7.0160 amu.

Ions differ from isotopes in that they deal with different numbers of electrons. An atom may lose or gain electrons from its neutral state. If an atom loses negatively charged electrons, it becomes known as a *cation* and bears a positive charge. I remember cation because the "t" reminds me of the "+" symbol in the positively charged ion. If an atom gains negatively charged electrons, it becomes an *anion* and bears a negative charge. I remember this because if you gain something negative, then you will have a negative charge.

The following equations illustrate how cations and anions are formed. Magnesium is a metal; metals tend to form cations by losing electrons. In equation 1, the neutral magnesium atom (Mg) loses two electrons to form the magnesium ion (Mg^{2+}). Nitrogen can form an anion by gaining electrons. Nonmetals, like nitrogen, tend to form anions. In equation 2, nitrogen gains three electrons to form the nitride anion.

$$(1)\ Mg \rightarrow Mg^{2+} + 2e^- \qquad (2)\ N + 3e^- \rightarrow N^{3-}$$

The compressed periodic table (no transition or inner transition metals are displayed) in **Table 3.2** illustrates some ion charge trends of the main group elements. The first group has a +1 charge; the second group has a +2 charge. The third main group has a +3 charge. The fifth and sixth main groups predominantly have a –3 and a –2 charge, respectively. The seventh main group has a –1 charge. This is a good trend to help you learn the charges of some common ions of elements.

TABLE 3.2 Common Ion Charge Trends

+1	+2		+3		−3	−2	−1	
hydrogen 1 H⁺								helium 2 He
lithium 3 Li⁺	beryllium 4 Be²⁺		boron 5 B	carbon 6 C	nitrogen 7 N³⁻	oxygen 8 O²⁻	fluorine 9 F⁻	neon 10 Ne
sodium 11 Na⁺	magnesium 12 Mg²⁺		aluminum 13 Al³⁺	silicon 14 Si	phosphorus 15 P³⁻	sulfur 16 S²⁻	chlorine 17 Cl⁻	argon 18 Ar
potassium 19 K⁺	calcium 20 Ca²⁺		gallium 31 Ga³⁻	germanium 32 Ge	arsenic 33 As (5+, 3+, 3−)	selenium 34 Se²⁻	bromine 35 Br⁻	krypton 36 Kr
rubidium 37 Rb⁻	strontium 38 Sr²⁻		indium 49 In³⁻	Sn⁴⁺ / Sn²⁺	Sb⁵⁺ / Sb³⁺	Tellurium 52 Te²⁻	iodine 53 I⁻	xenon 54 Xe
cesium 55 Cs⁺	barium 56 Ba²⁺		Tl³⁻ / Tl⁺	Pb⁴⁺ / Pb²⁺	Bi⁵⁺ / Bi³⁺	polonium 84 Po (4−, 2−, or 2−)	astatine 85 At⁻	radon 86 Rn
francium 87 Fr⁻	radium 88 Ra²⁻							

5. What type of elements (metals or nonmetals) form cations? Anions? What are the charges on the noble gases?

Review Questions

6. What are isotopes?

7. What are ions?

8. Compare and contrast isotopes and ions.

9. What happens if the number of protons in an atom changes? *This is neither an isotope nor an ion!*

Application Questions

10. Calculate the number of protons, neutrons, and electrons for the following elements.

    (a) ^{41}K

    (b) ^{63}Cu

    (c) ^{1}H

    (d) ^{17}O

11. Iron has four naturally occurring isotopes. They have 28, 30, 31, and 32 neutrons, respectively. Write the notation for each of the four isotopes.

12. Complete the following table.

Symbol	Mass #	Atomic #	Protons	Neutrons	Electrons
	36		18		
Ca^{2+}	40				
Cu				34	
		16		16	18

13. Europium has two isotopes. One of them is europium-153, with a mass of 152.9212 amu and an abundance of 52.19%. Calculate the isotopic mass of europium-151, the other isotope, which has a 47.81% abundance.

14. Thallium has two naturally occurring isotopes. Thallium-203 has a mass of 202.9723 amu, and thallium-205 has an isotope mass of 204.9744 amu. Calculate the percent abundance of each isotope.

15. Write the following elements with their common ionic charges.

   (a) Ca

   (b) P

   (c) I

   (d) Na

CHAPTER 4

Electronic Structure I

What is electromagnetic radiation?

How do atoms interact with light?

Learning Outcomes

- Apply the fundamental equations of light to calculate frequency, wavelength, and energy.
- Discuss the relationship between light and electrons.
- Explain the Bohr model of an atom.
- Calculate the energy related to electron transitions within a hydrogen atom.

Prerequisite Knowledge

- Chapter 3: Atomic Structure

Background Information and Questions

Much of what we know about the electrons in an atom's structure was discovered using light. Before we discuss electrons, let's talk about light. Light is known in the scientific world as *electromagnetic radiation*. Electromagnetic radiation is the flow of energy in the form of oscillating electric and magnetic fields. Those oscillating fields mean we can think of light as a wave, as shown in **Figure 4.1**.

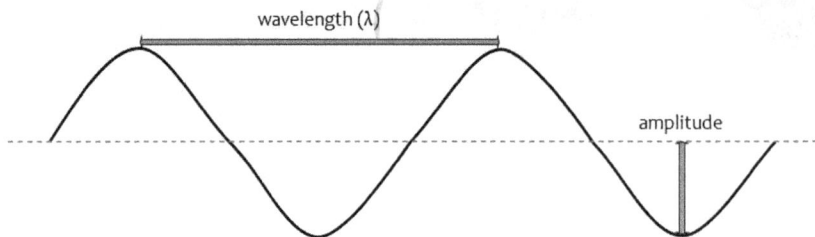

FIGURE 4.1 Wavelength and amplitude.

The *wavelength* of electromagnetic radiation is the distance from a point on one wave to the same point on the next wave. The *amplitude* is the maximum height of the wave, as measured from the center of the wave. By varying the wavelength, we can get different types of electromagnetic radiation. The seven main types of electromagnetic radiation are radio waves, microwaves, infrared radiation, visible light, ultraviolet (UV) light, X-rays, and gamma (γ) rays. These are displayed on the electromagnetic radiation spectrum in **Figure 4.2** in order of increasing wavelength from left to right.

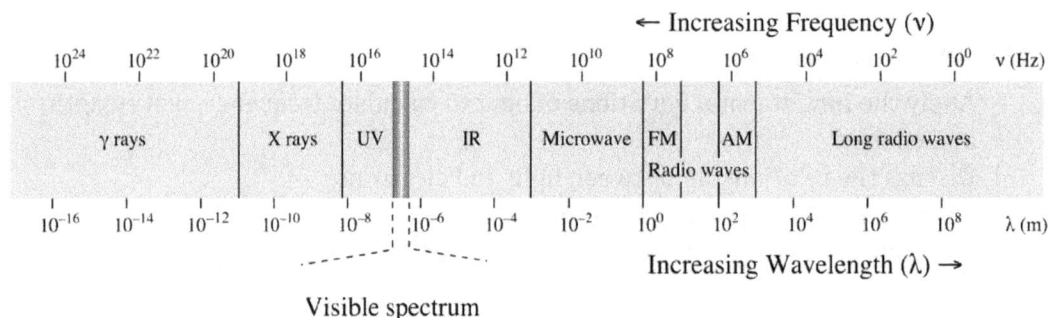

FIGURE 4.2 Electromagnetic spectrum.

1. Which would be more dangerous, electromagnetic radiation with a long wavelength or a short one? Why?

2. Why do you think UV radiation only gives you a sunburn, but X-rays can be used to determine bone structure?

Scientists have determined that electromagnetic radiation travels at 299 million meters per second, which is commonly rounded to 3.00×10^8 m/s. Because all light travels at the same speed, the frequency (ν) of wavelengths passing a point in space changes as the wavelength changes. In other words, a long wavelength will have a low frequency and vice versa. Wavelength and frequency are inversely proportional. Note the inverse proportionality in the electromagnetic spectrum in Figure 4.2. This relationship is modeled in the following equation:

$$c = \lambda\nu$$

c is the speed of light, λ is the wavelength, and ν is the frequency.

The frequency and wavelength can be related to the energy of the electromagnetic radiation. A long wavelength means a low frequency and low energy electromagnetic radiation. A short wavelength has a high frequency and high energy; therefore, energy is directly proportional to frequency and inversely proportional to wavelength. The relationships are represented by the following two equations:

$$E = h\nu \qquad E = \frac{hc}{\lambda}$$

E equals energy, and h is Planck's constant (6.626×10^{-34} J·s).

3. Which type of radiation will have the lowest energy?

4. How does its energy relate to the frequency and wavelength of the type of radiation in question 2?

Now that we've discussed light, let's talk about electrons. In 1905 Albert Einstein published the "Photoelectric Effect," which established that light and electrons can interact.[1] The findings of many other scientists, such as Louis de Broglie, supported Einstein's conclusions. De Broglie is credited with the *particle–wave duality*, meaning

---

1 Smith, N. (2019, November 6). Research Guides: Annus Mirabilis of Albert Einstein: Introduction. https://www.loc.gov/rr/scitech/SciRefGuides/einstein.html.

that an electron (particle) can behave like a light wave, and a light wave can behave like a particle. This idea is illustrated in the following equation:

$$\lambda = \frac{h}{mv}$$

m is mass, and v is velocity.

5. Describe the relationship between mass and wavelength in the equation for particle–wave duality.
    Hint: Is this an inverse or a direct relationship? That is, will they increase/decrease together, or does one go up as the other goes down?

The year before de Broglie's discovery, Niels Bohr concluded that electrons are located in successive energy levels around the nucleus. Using previous findings that energy can exist in quantized packets called *photons*, Bohr hypothesized that electrons in an atom exist on discrete energy levels around the nucleus, as shown in **Figure 4.3**.

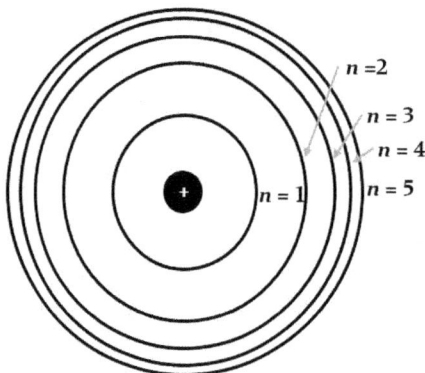

FIGURE 4.3 Bohr model of the atom.

The energy (E) of electrons can be calculated based on their energy level, as follows:

$$E = \frac{R_H}{n^2}$$

$R_H = 2.18 \times 10^{-18}$ J, and n = energy level.

One can also calculate the change in energy (DE) as an electron moves up or down energy levels:

$$\Delta E = -R_H\left(\frac{1}{n_f^2} - \frac{1}{n_i^2}\right)$$

R_H = 2.18 × 10^{-18} J, n_f = final energy level, and n_i = initial energy level.

When an electron moves up an energy level, it is referred to as *absorption*; the electron is absorbing energy to move to a higher energy level. When an electron moves to lower energy levels, it is referred to as *emission*, because it is emitting energy. When electrons reach a higher energy level, this is called the *excited state*. When the electrons are at their lowest energy level, this is called the *ground state*.

Bohr's model works only for hydrogen because it has only one electron. The energy calculations associated with multiple electrons are more complicated; they must account for repulsive forces and nuclear shielding.

6. Write a transition (i.e., n = 1 → n = 2) that would result in emission (negative value) of energy.

Review Questions

7. The relationships of the properties in electromagnetic radiation are as follows: the higher the _____, the lower the _____, and the lower the _____ _____.

8. How does light relate to electrons?

9. Without performing any calculations, rank the following transitions from the lowest to the highest emission of energy.

(a)   n = 2 → n = 1

(b)   n = 4 → n = 2

(c) $n = 5 \rightarrow n = 4$

(d) $n = 6 \rightarrow n = 5$

10. Describe the shortcomings of the Bohr model of the atom.

Application Questions

11. What is the frequency of light that has a frequency of 4.69×10^{-7} m?

12. What is the energy of that electromagnetic radiation described in question 11?

13. Determine the energy of blue light, which has a wavelength of 485 nm.

14. Determine the energy of a mole of photons of red light ($\lambda = 695$ nm).

15. What would be the velocity associated with an electron (mass = 9.11×10^{-31} kg) with a wavelength of 5.28×10^{-7} m?

16. Which of the following transitions would result in an absorption with the shortest wavelength?

    (a) $n = 1 \rightarrow n = 2$

    (b) $n = 2 \rightarrow n = 3$

    (c) $n = 3 \rightarrow n = 4$

    (d) $n = 4 \rightarrow n = 5$

17. Calculate the energy absorption as an electron moves from n = 3 to n = 6. What would be the wavelength of light necessary to excite the electron?

Figure Credit

Fig. 4.2: Copyright © by Philip Ronan (CC BY-SA 3.0) at https://commons.wikimedia.org/wiki/File:EM_spectrumrevised.png.

CHAPTER 5

Electronic Structure II

What are orbitals?

Where can we find certain electrons in the electron cloud?

Learning Outcomes

- Recognize the shapes and orientations of orbitals.
- Assign the four quantum numbers to any electron in an atom or ion.
- Utilize the Aufbau principle, Hund's rule, and the Pauli exclusion principle to write electron configurations of atoms and ions.
- Recognize and rationalize exceptions to predicted electron configurations.

Prerequisite Knowledge

- Chapter 3: Atomic Structure
- Chapter 4: Electronic Structure I

Background Information and Questions

Based on the work of famous scientists, such as Niels Bohr, Louis de Broglie, Werner Heisenberg, and many others, Erwin Schrödinger developed the field of quantum mechanics. From his investigations into quantum mechanics, we get quantum numbers. Four main quantum numbers allow us to identify areas where there is a high probability of finding an electron. Those quantum numbers are listed below.

- Principal quantum number (n) has integer values from 1 to ∞; for example, $n = 1, 2, 3, \ldots, \infty$.

- Angular momentum quantum number (l) has integer values from 0 to $n - 1$; for example, $l = 0, 1, 2, \ldots, n - 1$.
- Magnetic quantum number (m_l) has integer values of $-l$ to $+l$; for example, $m_l = -1, 0, 1$
- The magnetic spin quantum number (m_s) has integer values of $-1/2$ or $+1/2$.

The principal quantum number, angular momentum quantum number, and magnetic quantum number all describe orbitals. *Orbitals* are mathematical functions that define the space in which an electron exists with a high probability (95%). The principal quantum number relates to the size of the orbital(s); the angular momentum quantum number relates to the shape of the orbital; and the magnetic quantum number relates to the spatial orientation of the orbital. **Table 5.1** illustrates the relationship between the principal quantum number and the size of the orbital, and **Table 5.2** depicts the relationship between the orbital shape and the angular momentum quantum number.

1. If $n = 2$, what would be all the possible values of l?

2. If $n = 4$, what would be all the possible values of m_l?

3. Could an orbital exist that had values of $n = 2$, $l = 1$, and $m_l = -2$?

TABLE 5.1 Relationship Between the Principal Quantum Number and the Size of the Orbital

$n = 1$	$n = 2$	$n = 3$

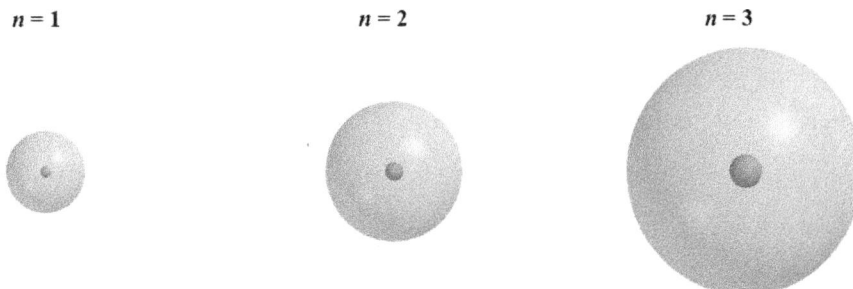

TABLE 5.2 Relationship Between Orbital Shape and the Angular Momentum Quantum Number

$l = 0$	$l = 1$			$l = 2$				
s orbital	p orbitals			d orbitals				
$m_l = 0$	$m_l = -1$	$m_l = 0$	$m_l = +1$	$m_l = -2$	$m_l = -1$	$m_l = 0$	$m_l = +1$	$m_l = +2$

4. Complete the following table.

n	l	Orbital Notation	m_l	Number of Orbitals
1	0			1
2	0	2s		
2	1		−1, 0, 1	
3		3s	0	1
3	1	3p		3
3		3d	−2, −1, 0, 1, 2	
4	0		0	1
4	1	4p	−1, 0, 1	
4	2			5
4		4f		7
5	0		0	
5		5p		3
5	2		−2, −1, 0, 1, 2	5
5	3		−3, −2, −1, 0, 1, 2, 3	

5. Which quantum number(s) relate only to orbitals? Which quantum number(s) relate only to electrons?

With the quantum numbers and the discoveries of Heisenberg, Schrödinger, Bohr, and many others, we can map out the locations of electrons in an atom. We can summarize the quantum numbers and electron locations by writing electron configurations of atoms and ions. Heisenberg's uncertainty principle states we can never know the exact position and momentum of an electron; instead, we can plot the probability of finding one. Orbitals are the probability plots that indicate areas where electrons might reside in the electron cloud of an atom. Before writing electron configurations, let's talk about some rules and principles that we use to fill orbitals with electrons.

▪ *Aufbau principle*: States that in the ground state, electrons fill the lowest energy orbitals first (translation: we fill atoms by increasing atomic number).

▪ *Hund's rule*: States that the lowest energy state of degenerate orbitals in a sub-shell is the maximum number of parallel spins (translation: orbitals of the same energy will half fill first, due to having to overcome pairing energy).

▪ *Pauli exclusion principle*: States that no two electrons can have the same set of quantum numbers (translation: in an orbital, which can hold two electrons, each electron must have a different spin).

When writing the electron configuration of an atom, we always start at the lowest energy orbital, the 1s orbital; then electrons are added to successive, higher energy orbitals: 2s, 2p, and so on. **Figures 5.1** and **5.2** detail which orbitals come next as you add electrons to atoms. Figure 5.1 illustrates the order in which to fill the orbitals, whereas Figure 5.2 shows the relative energies of the orbitals and the electron configuration of bromine. This type of configuration is known as *orbital box notation* and will be discussed in greater detail later in this chapter.

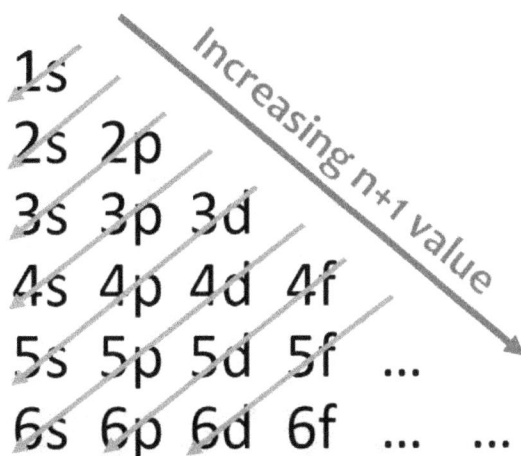

FIGURE 5.1 Order in which to fill orbitals.

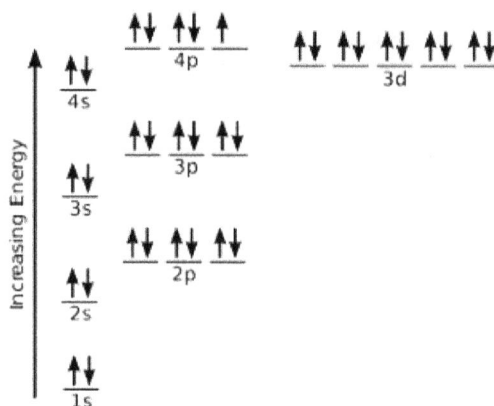

FIGURE 5.2 Relative energies of orbitals and electron configuration of bromine.

Another way of visualizing the orbitals is the position of the orbital block on the periodic table (**Table 5.3**). The orbital blocks represent the locations of the outermost electrons (called *valence electrons*) for each atom. For example, the outermost electron on a boron atom is in the 2p shell.

TABLE 5.3 Periodic Table With Orbital Blocks

6. How many electrons would exist for the principal quantum number $n = 4$?

Let's write the electron configuration of chlorine three ways: the full configuration, the noble gas configuration, and the orbital box notation.

To write the full configuration, start at hydrogen on a periodic table and work through the increasing atomic numbers and the orbital blocks atom by atom until reaching the desired element. The configurations for the first several elements are written out below.

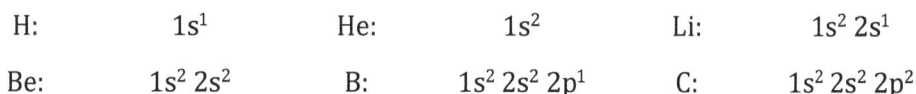

H:	$1s^1$	He:	$1s^2$	Li:	$1s^2\,2s^1$
Be:	$1s^2\,2s^2$	B:	$1s^2\,2s^2\,2p^1$	C:	$1s^2\,2s^2\,2p^2$

In the above configurations the exponent represents the number of electrons in the listed orbital. For example, lithium has 2 electrons in the 1s orbital and 1 electron in the 2s orbital, making it $1s^2\,2s^1$.

Notice that the s-block only has two elements for each energy level. That's because there is one s-orbital per energy level, which can hold only two electrons. The p-block has six elements because there are three p-orbitals, one oriented along each three-dimensional axis: p_x, p_y, and p_z. If each p-orbital can hold two electrons, then the p-block can have a total of six electrons per energy level.

Now, let us write the full electron configuration of chlorine. Working through all of the orbital blocks (as above), we eventually arrive at chlorine, which has a configuration of $3p^5$. The full electron configuration includes all the electrons prior to $3p^5$ (**Figure 5.3**). Let's check to make sure we are correct; chlorine has 17 electrons. If we add the exponents in our configuration, we get 17. We have written the correct full electron configuration. Note the labeling of valence electron and core electrons (the inner electrons of an atom).

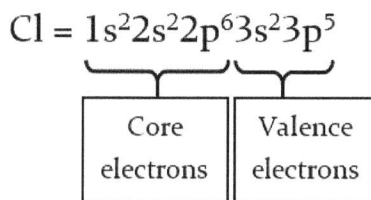

$$Cl = 1s^2 2s^2 2p^6 3s^2 3p^5$$

Core electrons	Valence electrons

FIGURE 5.3 Electron configuration of chlorine.

7. What would be the full configuration for neon?

The noble gas notation is a shortcut to the full electron configuration. To write the noble gas notation, start at chlorine and go back (decreasing atomic numbers) to the previous noble gas. Write that noble gas in brackets, then proceed forward (increasing atomic numbers) to chlorine, writing the electron configuration as you go (**Figure 5.4**). The noble gas in brackets represents the core electron configuration. This type of configuration is often employed to highlight the valence shell of an atom or to write the electron configuration of very large elements because the full electron configuration would be quite long due to the high number of electrons.

$$Cl = [Ne]3s^2 3p^5$$

FIGURE 5.4 Noble gas notation of chlorine.

The orbital box notation features boxes or lines for each orbital, and each electron is represented by an arrow. Arrows are drawn up or down to represent electron spin. Orbital box notation can be written out in a linear fashion (**Figure 5.5**) or representing the relative energies (**Figure 5.6**).

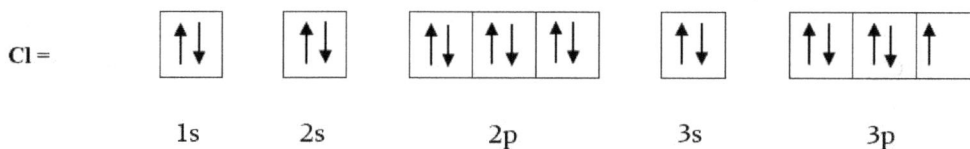

FIGURE 5.5 Orbital box notation: linear style.

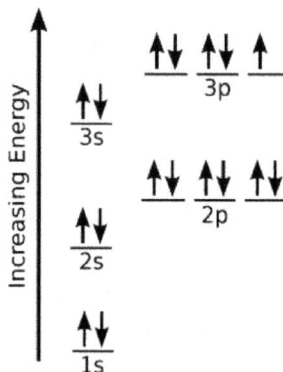

FIGURE 5.6 Orbital box notation: relative energies.

Some elements in the d and f blocks have electron configurations that are exceptions. Copper is one of those elements. Its electron configuration would appear to be $[Ar]4s^2 3d^9$. However, the 3d shell would prefer to be fully filled, so the actual configuration is [Ar]$4s^1 3d^{10}$, shown in orbital box notation in **Figure 5.7**. This occurs because a fully filled 3d orbital is more stable, energetically, than a partially filled orbital.

Cu = [Ar]

$\uparrow$ $\uparrow\downarrow$ $\uparrow\downarrow$ $\uparrow\downarrow$ $\uparrow\downarrow$ $\uparrow\downarrow$

4s   3d

FIGURE 5.7 Orbital box notation for copper.

8. Write the electron configuration for silver using noble gas notation.
   (Hint: Silver is right beneath copper on the periodic table!)

9. Draw the orbital box configuration for silver (Ag) below.

Review Questions

10. In your own words, summarize the rules of the four quantum numbers.

11. Using orbital box notation, illustrate violations of the following.

    (a) Aufbau principle

    (b) Hund's rule

(c) Pauli exclusion principle

12. If n = 3 and l = 1, how many values for m_l are possible?

Application Questions

13. Sketch the shape of the orbital with the following set of quantum numbers: $n = 3$, $l = 2$, and $m_l = 0$

14. How many electrons exist in the 4s orbital of an Ar atom?

15. Write the full electron configuration and the noble gas electron configuration of oxygen.

16. What element has the noble gas electron configuration $[Kr]5s^2 4d^7$?

17. Write the electron configuration (both noble gas and full) for each of the following atoms or ions.

    (a) Mg

    (b) Sr^{2+}

    (c) S^{2-}

    (d) Ag

    (e) Cr

    (f) Ni^{2+}

Figure Credits

CHAPTER 6

Chemical Bonding

What holds two atoms together?

How do bonds give rise to certain properties?

How do bonds cause a reaction to be endothermic or exothermic?

Learning Outcomes

▨ Identify elements commonly found in ionic, covalent, or metallic bonds.
▨ Describe the formation of ionic or covalent bonds.
▨ Define electronegativity and determine the polarity of a covalent bond.
▨ Calculate the enthalpy of a reaction using bond dissociation energies.

Prerequisite Knowledge

▨ Chapter 4: Electronic Structure I
▨ Chapter 5: Electronic Structure II

Background Information and Questions

Chemical bonds are the result of attractive and repulsive forces between two atoms, and bonds form molecules. The combination of different elements results in different bond types. This chapter will evaluate three main bond types: ionic, covalent, and metallic.

Ionic bonds are the electrostatic forces between a metal atom (usually a cation) and a nonmetal atom (usually an anion). During the formation of an ionic bond, one atom or molecule gains an electron(s) from the other atom or molecule, so one species transfers an electron(s) to another. Once the electron is transferred, the atom or molecule that lost the electron(s) has an overall positive charge (it is a cation), whereas the atom or molecule that gained the electron(s) now has an overall negative charge (it is an anion). The attractive forces between the now formed ions are *electrostatic*, meaning they are held together by their opposite charges. Repulsive forces are also keeping the atoms apart; these forces include the repulsion of the electrons ("like" charges repel) and the positively charged nuclei. **Figure 6.1** illustrates the formation of an ionic bond between sodium and chlorine. Notice how the sodium loses one electron and obtains a noble gas electron configuration, while the chlorine atom gains that electron to gain the noble gas configuration. This forms the sodium ion and the chloride ion, which results in the ionic compound sodium chloride.

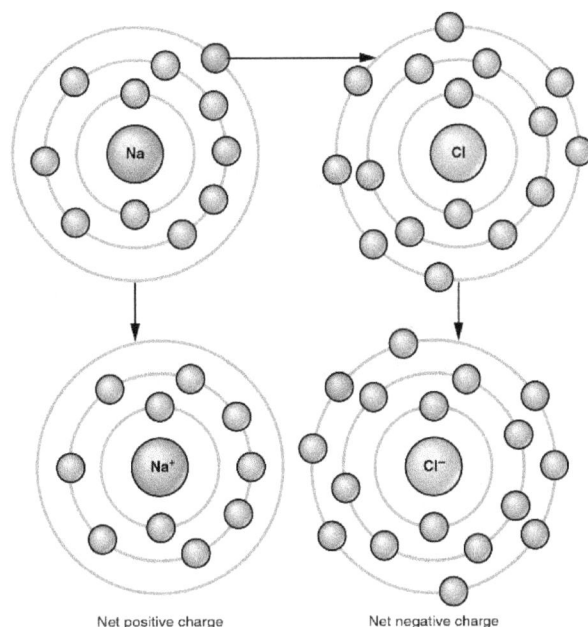

Net positive charge   Net negative charge

FIGURE 6.1 Formation of an ionic bond between sodium and chlorine.

1. What charge would you expect the following elements to have when they become ions?

   Al   Se   P   K   I   Ca

Transferring electrons from one atom to another is not the only way bonds can be formed. Covalent bonds often are formed between two nonmetal atoms; they involve the *sharing* of two valence electrons between atoms. The attractive force holding the atoms together is the interaction of the shared electrons with both nuclei and the overlap of the orbitals of each atom. The repulsive forces are the same as experienced in an ionic bond: repulsion of other electrons and repulsion of the positive nuclei. **Figure 6.2** illustrates the formation of a covalent bond between two hydrogen atoms. Notice that each hydrogen nuclei has a +1 charge, representing its one positively charged proton. Also, notice that each hydrogen atom has one electron and contributes that one electron to the bond. In a covalent bond, each atom contributes one electron.

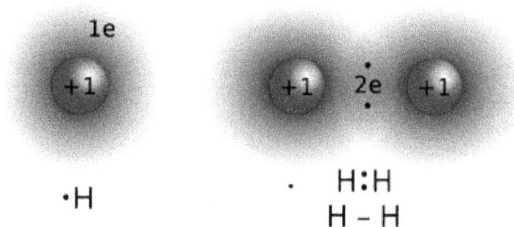

FIGURE 6.2 Formation of a covalent bond between two hydrogen atoms.

Some covalent bonds share electrons equally, whereas others do not. A covalent bond that does not share electrons equally is called a *polar covalent bond*. This unequal sharing is due to differences in electronegativity of the atoms in the bond. Electronegativity (χ) is the ability of an atom to attract electrons to itself. The periodic table in **Figure 6.3** shows the electronegativity of each element on the Pauling scale.

Group (vertical) / Period (horizontal)	1	2	3	4	5	6	7	8	9	10	11	12	13	14	15	16	17	18
1	H 2.20																	He
2	Li 0.98	Be 1.57											B 2.04	C 2.55	N 3.04	O 3.44	F 3.98	Ne
3	Na 0.93	Mg 1.31											Al 1.61	Si 1.90	P 2.19	S 2.58	Cl 3.16	Ar
4	K 0.82	Ca 1.00	Sc 1.36	Ti 1.54	V 1.63	Cr 1.66	Mn 1.55	Fe 1.83	Co 1.88	Ni 1.91	Cu 1.90	Zn 1.65	Ga 1.81	Ge 2.01	As 2.18	Se 2.55	Br 2.96	Kr 3.00
5	Rb 0.82	Sr 0.95	Y 1.22	Zr 1.33	Nb 1.6	Mo 2.16	Tc 1.9	Ru 2.2	Rh 2.28	Pd 2.20	Ag 1.93	Cd 1.69	In 1.78	Sn 1.96	Sb 2.05	Te 2.1	I 2.66	Xe 2.60
6	Cs 0.79	Ba 0.89	*	Hf 1.3	Ta 1.5	W 2.36	Re 1.9	Os 2.2	Ir 2.20	Pt 2.28	Au 2.54	Hg 2.00	Tl 1.62	Pb 2.33	Bi 2.02	Po 2.0	At 2.2	Rn 2.2
7	Fr 0.7	Ra 0.9	**	Rf	Db	Sg	Bh	Hs	Mt	Ds	Rg	Uub	Uut	Uuq	Uup	Uuh	Uus	Uuo
Lanthanides	*	La 1.1	Ce 1.12	Pr 1.13	Nd 1.14	Pm 1.13	Sm 1.17	Eu 1.2	Gd 1.2	Tb 1.1	Dy 1.22	Ho 1.23	Er 1.24	Tm 1.25	Yb 1.1	Lu 1.27		
Actinides	**	Ac 1.1	Th 1.3	Pa 1.5	U 1.38	Np 1.36	Pu 1.28	Am 1.13	Cm 1.28	Bk 1.3	Cf 1.3	Es 1.3	Fm 1.3	Md 1.3	No 1.3	Lr 1.291		

Periodic table of electronegativity using the Pauling scale

FIGURE 6.3 Periodic table of electronegativity using the Pauling scale.

To determine if a bond is polar or not polar (which is called *nonpolar*), one must determine the differences in electronegativity of the two elements involved in the bond. If the difference in electronegativity is 0.0 to 0.5, then the bond is considered a nonpolar covalent bond. This means there is equal or almost equal sharing of the two electrons in the bond. If the difference is 0.5 to 1.5, the bond is considered a polar covalent bond. If the difference is greater than 1.5, then the bond is considered ionic, meaning that there is an electron transfer, not sharing. In terms of ionic bonds, note that elements that form anions have higher electronegativities, and those that tend to form cations are metals with lower electronegativities. This leads to that large difference in electronegativity. In the previous two paragraphs, this book has separated covalent and ionic bonds; in actuality, most bonds are a mixture of ionic and covalent character based on their electronegativity differences.

When a covalent bond is polar, we draw a dipole moment to indicate the polarity of the bond. In **Figure 6.4**, you can see a polar bond with a dipole moment; the dipole moment is drawn pointing toward the more electronegative element.

$$\delta+ \quad \longrightarrow \quad \delta-$$
$$\text{H} - \text{F}$$

FIGURE 6.4 Polar bond with a dipole moment.

Note: Physicists and the International Union of Pure and Applied Chemistry (IUPAC) draw the dipole moment in the opposite direction, with the dipole pointing toward the partially positive end of the molecule. That end of the molecule will be attracted to an electrical charge if it is applied to the molecule.

2. Draw the dipole moments on the following bonds. If there is no dipole, write "nonpolar" beneath the molecule.

C–O   N–O   Cl–Cl

In a metal, the atoms are uniformly arranged, and metallic bonds are formed by the attraction of each metal's nucleus to delocalized electrons moving freely between uniformly arranged atoms. The repulsive forces are the same as experienced in an ionic or covalent bond: repulsion of other electrons and repulsion of the positive nuclei. Metallic bonding is often described as metal atoms floating in a "sea" of mobile electrons. This unique bonding between metal atoms gives metals unique properties; mobile electrons make metals very good conductors of heat and electricity. **Table 6.1** shows the "sea-of-electrons" model at the microscopic level for metallic bonding and summarizes the other two types of bonds that you have been reading about.

TABLE 6.1 Types of Bonds

	Covalent	**Ionic**	**Metallic**
Symbolic	Cl_2 (l)	NaCl (s)	Au (s)
Microscopic			
Macroscopic			
	Chlorine	Salt Granule	Gold Ring

3. Bonding explains why certain compounds have unique properties. Why do solutions of ionic compounds and metals tend to be conductors of electricity while molecules with covalent bonds tend to be insulators?

Bonds are energy! They are combinations of attractive and repulsive forces that release energy when formed (*exothermic*) and require energy to break (*endothermic*). The potential energy well shown in **Figure 6.5** illustrates how the attractive and repulsive forces control the bond length in covalent bonds. Let's move from right to left along the *x*-axis, which is the distance between atoms. As atoms are infinitely far apart, there are no interactions, so there is no energy released as the bond forms. As the atoms move closer together, there is a greater attractive force than repulsive, and more and more energy is released. When the atoms are too close together, the repulsive forces are greater than the attractive ones, so bond formation quickly becomes endothermic. At the energy minimum, the bond is most stable at that internuclear distance. This distance is known as the *bond length*.

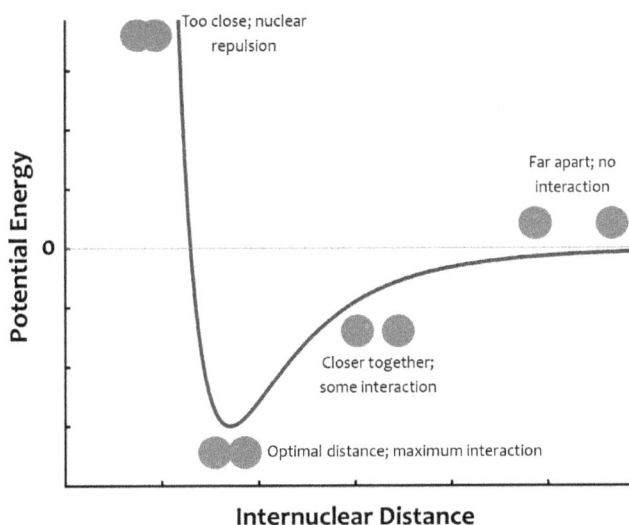

FIGURE 6.5 Relationship between potential energy and internuclear distance.

The difference between the energy at the lowest point on the potential well (maximum interaction) and the energy when the atoms are infinitely separated (no interaction) is called the *bond dissociation energy*. Bond dissociation energy is associated with a covalent bond's length and bond order. Bond order is the number of electron pairs shared between two atoms; a single bond has a bond order of one, a double bond has an order of two, and a triple bond has a bond order of three. Resonance structures have decimal bond orders due to the delocalized electrons. The general trends are that a shorter bond length and a higher bond order result in a higher bond dissociation energy.

4. Why does the potential energy go up if nuclei get too close together? What does this abrupt rise in the potential energy mean for bond formation?

5. Given that covalent bonds form from the overlap of two atomic orbitals and the sharing of valence electrons, how do you think atomic radius is related to bond length?

Bond dissociation energy can be used to calculate the enthalpy of reactions. Enthalpy is a common way that scientists report the heat energy associated with a chemical reaction and will be covered in more detail in a later chapter (Chapter 30: Energetics II). To calculate an enthalpy of reaction from bond dissociation energies, you must subtract the bond dissociation energies of bonds formed from the bond dissociation energies of bonds broken.

$$\Delta H_r = bonds\ broken - bonds\ formed$$

Using the reaction below and **Table 6.2** of bond dissociation energies, we can calculate the enthalpy of the reaction. We must multiply each bond dissociation energy by the number of that bond that appears in the reaction.

The example shown below uses bond dissociation energies to calculate the enthalpy of reaction for the provided reaction.

$$H_2 + Cl_2 \rightarrow 2HCl$$

TABLE 6.2 Average Bond Dissociation Energies (kJ/mol)

Bond	Bond Dissociation Energy (kJ/mol)
H–H	436
Cl–Cl	243
H–Cl	432
C–H	410
C–C	350
C–O	350
C = O	732
C ≡ O	1,077
O = O	498
O–O	180
O–H	460

$$\Delta H_r = bonds\ broken - bonds\ formed$$
$$\Delta H_r = \left((H-H) + (Cl-Cl) \right) - (2*H-Cl)$$
$$\Delta H_r = (436 + 243) - (2 \times 432)$$
$$\Delta H_r = -185\ kJ$$

The enthalpy of reaction for the reaction of hydrogen with chlorine is therefore −185 kilojoules, which is exothermic as the enthalpy/heat is released (the answer has a negative value and would be considered a product).

Review Questions

6. Explain the relationship of electronegativity with covalent and ionic bonds.

7. How do bond dissociation energies allow for the calculation of reaction enthalpies?

Application Questions

8. Predict which type of bond you would expect in the following molecules.

   NaF   AgCl BH_3

   CO_2 PCl_3 Na_3PO_4

9. For each of the following pairs, circle the bond you would expect to be the most polar.

   C – F or H – Cl   Si – Cl or P – O   C – N or C – O

10. The Lewis dot structures for methanol (CH_4O), carbon dioxide (CO_2), and carbon monoxide (CO) are provided below. Calculate the bond order and rank each molecule in terms of carbon–oxygen bond strength.

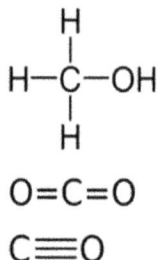

$$
\begin{array}{c}
\text{H} \\
| \\
\text{H—C—OH} \\
| \\
\text{H}
\end{array}
$$

$$\text{O=C=O}$$

$$\text{C}\equiv\text{O}$$

11. Acetic acid is formed by carbonylation of methanol, as shown below. Using the table of bond dissociation energies above, calculate the enthalpy of the reaction.

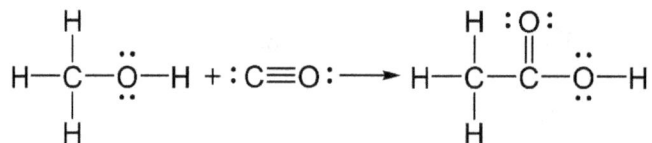

$$
\text{H—C—O—H} + \text{:C}\equiv\text{O:} \longrightarrow \text{H—C—C—O—H}
$$

12. Using bond dissociation energies from Table 6.2, estimate the dissociation energy of the C = O bond in CO_2 (g).

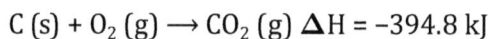

$$\text{C (s)} + O_2 \text{ (g)} \longrightarrow CO_2 \text{ (g)} \quad \Delta H = -394.8 \text{ kJ}$$

Figure Credits

CHAPTER 7

Lewis Dot Structures

How do we represent covalent molecules?

What is the bonding environment of atoms in covalent molecules?

Learning Outcomes

- Draw Lewis dot structures of small molecules that obey or violate the octet rule.
- Define resonance, and recognize it in Lewis dot structures.
- Utilize formal charges to determine the most correct Lewis dot structures.

Prerequisite Knowledge

- Chapter 5: Electronic Structure II
- Chapter 6: Chemical Bonding

Background Information and Questions

Lewis dot structures provide chemists with important information about the bonding environment of atoms in covalent molecules. These two-dimensional representations of covalent molecules were developed by G. N. Lewis. Before we learn to draw Lewis dot structures, let's learn about Lewis dot symbols. The dots represent the number of valence electrons for an atom.

TABLE 7.1 Relationship Between Group Number, Electron Configuration, and Lewis Dot Symbol

Group 1A	Group 2A	Group 3A	Group 4A	Group 5A	Group 6A	Group 7A	Group 8A
$[He]2s^1$	$[He]2s^2$	$[He]2s^22p^1$	$[He]2s^22p^2$	$[He]2s^22p^3$	$[He]2s^22p^4$	$[He]2s^22p^5$	$[He]2s^22p^6$
Li·	Be·	·B·	·C·	·N·	·O·	·F:	:Ne:

Table 7.1 relates the group number, electron configuration, and Lewis dot symbol. The electron configurations only display the valence electrons, which are the outermost electrons; notice how the sum of exponents aligns with the group number and the dots on the Lewis dot symbols. Lewis dot symbols give us insight into the bonding environment of atoms with certain numbers of valence electrons. For example, carbon has four valence electrons but would like to attain eight electrons (the valence shell of a noble gas). Carbon can get eight valence electrons by sharing four more electrons with other atoms through covalent bonds. A covalent bond is two electrons; therefore, carbon can potentially form four bonds, which will give it the eight valence electrons it needs.

1. How many bonds do you think fluorine might form with other atoms?

Figure 7.1 is the Lewis dot structure of hydrochloric acid. Electrons are treated as pairs in Lewis dot structures; they can be bonded pairs or lone pairs. Bonded pairs form covalent bonds; they are shared between atoms. Lone pairs are not bonded electrons.

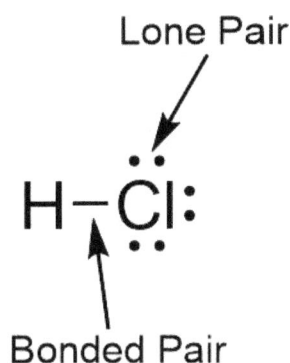

FIGURE 7.1 Lewis dot structure of hydrochloric acid.

2. Why do you think we draw Lewis dot structures for covalent molecules but not for ionic compounds?

Here are the rules for drawing Lewis dot structures:

I. Determine the total number of valence electrons in a molecule or ion. Remember to add valence electrons for anions and to subtract electrons for cations.

II. Establish which atom will be the molecule's central atom. Some molecules may have more than one central atom. Carbon is always a central atom; hydrogen is never a central atom.

III. Draw a skeletal structure with that atom as a central atom and evenly place the terminal atoms around the central atom.

IV. Add single bonds between the central atom(s) and the terminal atom(s).

V. Complete the octets of the terminal atoms by adding electrons as lone pairs. Remember that hydrogen does not follow the octet rule; it only requires a duet.

VI. Count the number of valence electrons in your structure. If you have remaining valence electrons in your total, then add those as lone pairs to the central atom. If you have no remaining valence electrons and your central atom does not have an octet, form double or triple bonds as necessary to complete the octet of the central atom(s).

> When drawing Lewis dot structures, charged molecules are always enclosed with brackets. The charge is then written on the outside of the brackets at the top right.

Let's draw a Lewis dot structure together. Let's draw ammonia, NH_3. Each number corresponds to the numbered rules above.

I. Ammonia has eight total valence electrons. Nitrogen has five valence electrons. Hydrogen has one; however, the ammonia has three hydrogens. Three hydrogens with one valence electron each contribute three valence electrons. To get the total, five plus three gives the molecule a total of eight valence electrons.

II. Nitrogen will be the central atom, since hydrogen is never a central atom.

III. We draw a skeletal structure with the nitrogen atom as the central atom and evenly place the hydrogen atoms around the central nitrogen.

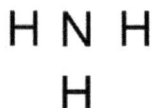

$$H \quad N \quad H$$
$$H$$

IV. We add single bonds between the central nitrogen and the terminal hydrogens.

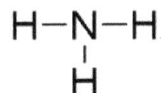

$$H-N-H$$
$$|$$
$$H$$

V. There is no need to complete the octets of the hydrogens; each only requires two electrons. The hydrogens in the structure in step IV already have two electrons that are being shared in the bonded pair with nitrogen.

VI. We have used only six of the total eight valence electrons; let us add the final two electrons to the nitrogen.

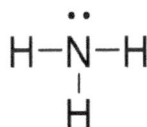

$$H-\ddot{N}-H$$
$$|$$
$$H$$

3. Which atom in the structure above satisfies the octet rule?

Once the Lewis structure is complete, always count to make sure you've used all the valence electrons and to make sure each atom has an octet or duet with regard to hydrogen. Another way to check your Lewis structure is to calculate formal charges. The *formal charge* is an integral charge assigned to an atom in a molecule. It's a simple way of determining where electron-rich or electron-poor areas are within a Lewis dot structure. Formal charge is determined via the following formula:

$$Formal\ charge\ =\ \#ve^- - ue^- - \frac{1}{2}be^-$$

whereas ve^- is valence electrons, ue^- is unpaired electrons, and be^- is bonded electrons.

The formula can be applied to a molecule by drawing a circle around the atom in which you intend to calculate the formal charge. When drawing the circle, cut the bonds to that atom in half. Consider the example for carbonate in **Figure 7.2**.

Once the intended atom is circled, count the number of electrons inside the circle. Remember that half of a bond is one electron. The top left oxygen on the carbonate molecule has seven electrons inside the circle. Now, subtract that number from the number of valence electrons of an oxygen atom. Oxygen has six valence electrons; six minus seven would give that oxygen a formal charge of negative one. **Figure 7.3** shows all of the formal charges for the carbonate molecule. Notice how the formal charges add to the overall charge on the carbonate molecule.

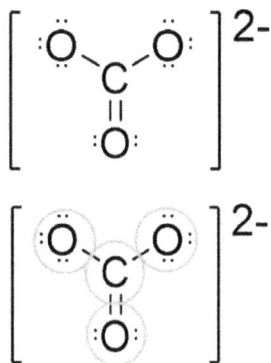

FIGURE 7.2 Circling of atoms in order to determine the formal charge.

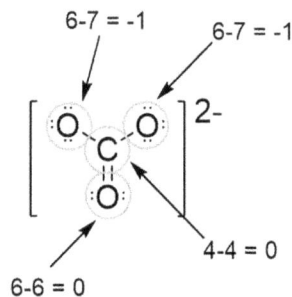

FIGURE 7.3 Determination of the formal charge of carbonate.

Formal charges can be used to determine whether you've drawn the most correct Lewis dot structure. For example, some molecules do not obey the octet rule. Only the central atoms in period 2 of the periodic table follow the octet rule. Any central atom that lies in period 3 or greater can have more than eight valence electrons. We can use formal charge to determine when those central atoms may violate the octet rule. The rules for utilizing formal charges to determine the most correct Lewis dot structure are as follows:

I. Formal charges in a Lewis dot structure should total to the overall charge on the molecule.

II. The Lewis dot structure with the lowest formal charges is the more stable Lewis structure.

III. If there must be a formal charge on an atom, the most stable structure has negative formal charges on the more electronegative atom and positive formal charges on the less electronegative atom.

4. Give an example of a situation where formal charge might take precedence over the octet rule.

A phenomenon in covalent bonding is electrons that are not localized between two atoms. These electrons are known as *delocalized electrons* and move between three or more atoms. This is known as *resonance*. Let us continue to use carbonate as an example. Recognizing resonance is important because it contributes to a molecule's properties. Resonance occurs when there is a double bond that can be drawn in multiple locations and still fit the rules of Lewis dot structures. The carbonate structures in **Figure 7.4** illustrate how the double bond can be drawn between the carbon and all three oxygens. In molecules that possess resonance, one must draw all possible resonance structures.

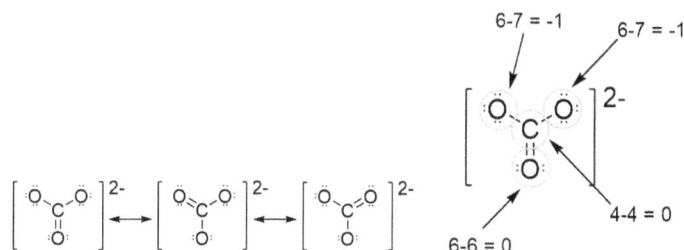

FIGURE 7.4 Resonance of carbonate.

The two π electrons in the double bond are delocalized over the entire carbonate molecule. There is not actually a double bond in a carbonate molecule. Imagine if the double bond were divided between all three single C–O bonds. Instead of two single bonds and a double bond, there are actually three single bonds and a third of a bond between the C and each O atom. Lewis dot structure rules do not allow for a 1.33 bond. Therefore, we must draw all three structures of carbonate to illustrate how the π electrons in the double bond are delocalized over the entirety of the molecule.

  5. Explain why NH_3 would not have resonance.

Review Questions

  6. How do Lewis symbols relate to Lewis structures?

  7. Why would you use formal charge?

8. Explain why one must draw all resonance structures of a molecule.

Application Questions

9. Write the Lewis symbol for boron. How many bonds do you think boron likes to form with other atoms?

10. Identify three possible elements that match the following Lewis symbols.

    (a)  $\overset{\bullet}{X}\cdot$

    (b)  $\cdot\overset{\bullet}{\underset{\bullet\bullet}{X}}\cdot$

    (c) 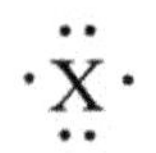 $\cdot\overset{\bullet\bullet}{\underset{\bullet\bullet}{X}}\cdot$

11. Draw the Lewis dot structure for the following molecules.

    (a) CH_2Cl_2

    (b) NO_3^-

(c) N_2

(d) $C_2H_4Cl_2$

12. Draw the MOST correct Lewis dot structure for the following molecules.
    Hint: Use formal charges!

    (a) SCN^-

    (b) SO_2

    (c) XeF_4

13. Identify the structure(s) that have resonance. Draw all the possible resonance structures of those molecules.

    (a) PF_3

    (b) SCN^-

(c) SO_4^{2-}

(d) H_2O

(e) CO_2

CHAPTER 8

Bonding Theory

Why and how do atoms bond?

Learning Outcomes

- Rationalize the formation of covalent bonds through orbital overlap.
- Compare and contrast σ and π bonds, and identify them in a molecular structure.
- Describe the concept of orbital hybridization, and determine the hybridization of an atom.
- Identify the HOMO and LUMO, and calculate the bond order using a molecular orbital diagram.

Prerequisite Knowledge

- Chapter 5: Electronic Structure II
- Chapter 6: Chemical Bonding

Background Information and Questions

The two covalent bonding theories in chemistry are valence bond theory and molecular orbital theory. Do not think of these as competing theories but rather as complementary theories. It is important to note that these two bonding theories apply to covalent bonds, not ionic or metallic bonds.

Valence bond theory theorizes that bonds form from the overlap of half-filled atomic orbitals. The new overlapping atomic orbitals are primarily localized between the two bonding atoms. Each atomic orbital is half filled, so it contributes one valence electron.

Remember that a covalent bond contains two valence electrons. Valence bond theory identifies two types of orbital overlap or bonds: sigma (σ) bonds and pi (π) bonds.

Recall from Chapter 6 that bonds are energy and that they have an optimal distance, which is identified by the potential energy minimum. These ideas still apply, with the addition that valence bond theory explains how orbitals overlap to form bonds. Sigma (σ) bonds are formed from the overlap of two half-filled s orbitals, an s and p orbital, or two end-to-end p orbitals. Pi (π) bonds form from the side-to-side overlap of two half-filled p orbitals, a p and d orbital, or two d orbitals. The σ and π bonds are depicted in **Figure 8.1**.

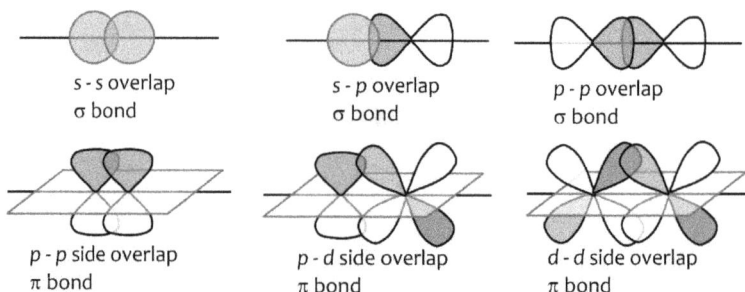

s - s overlap	s - p overlap	p - p overlap
σ bond	σ bond	σ bond
p - p side overlap	p - d side overlap	d - d side overlap
π bond	π bond	π bond

FIGURE 8.1 Formation of σ and π bonds.

1. Compare and contrast σ and π bonds.

2. In which orbitals are electrons located for each type of bond?

Whereas valence bond theory explains how bonds form, it does not provide an explanation for molecular shapes. To help describe how molecules get their three-dimensional shape, I will use methane, CH_4. Methane has a central carbon atom and a tetrahedral molecular geometry. If we think about the electron configuration of the central carbon atom ($1s^2 2s^2 2p^2$), there are only two unpaired valence electrons: the two electrons in the 2p orbitals. If we follow the convention of the valence bond theory, then carbon can only form two bonds. We know this is not true; carbon can form as many as four single bonds. We also know that carbon's four bonds are 109.5° apart, but the p orbitals that contain the valence electrons are 90° apart because they are located along each axis ($x, y,$ and z). The question is: "How does carbon form four bonds at 109.5°?" As illustrated by its electron configuration and valence bond theory, it is supposed to form two bonds that are 90° apart.

It is here that we introduce the idea of hybridization in valence bond theory. We can take two atomic orbitals and mix them to form new orbitals that can explain the molecular shapes that we see. Let's use H_2O in the diagram below to explain hybridization. Oxygen, the central atom, has an electron configuration of $1s^2 2s^2 2p^4$. Looking at the valence electrons in the 2s and 2p orbitals, we can see there are two unpaired electrons, meaning oxygen can form two bonds and have two lone pairs. However, the lone pairs would not be equal, because the 2s pair of electrons is lower in energy than the 2p pair. If we mix or hybridize the 2s and the three 2p orbitals, we can form four new orbitals that are equal in energy. Now, our lone pairs are equal in energy and our two unpaired electrons are equal and available to bond in a σ bond to the 1s orbital (blue sphere in **Figure 8.2**) of the hydrogen atoms. Because we mixed one s orbital and three p orbitals, we call this an sp^3 hybridization. These four hybrid orbitals arrange themselves as far from each other as possible around the central atom; in this way valence bond theory complements valence shell electron pair repulsion (VSEPR).

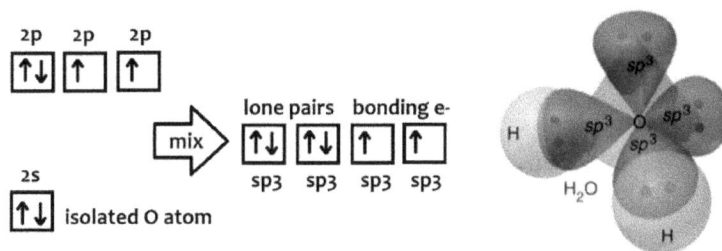

FIGURE 8.2 Hybrid orbitals.

Luckily, you won't have to think about the electron configuration of the central atom each time to determine the hybridization. **Table 8.1** relates the number of electron domains, electron pair geometry (EPG), and the hybridization of the central atom. When you are determining the hybridization of a central atom, just consider its EPG.

TABLE 8.1 Determination of Hybridization

Number of Electron Domains	Electron Pair Geometry	Hybridization
2	Linear	sp
3	Trigonal planar	sp^2
4	Tetrahedral	sp^3
5	Trigonal bipyramidal	sp^3d
6	Octahedral	sp^3d^2

3. What would the hybridization be if one mixed three p orbitals, one s orbital, and two d orbitals?

Valence bond theory explains multiple bonds by combining σ and π bonds. For example, a double bond is one σ and one π bond, and a triple bond is formed from one σ and two π bonds. To explain how a double bond works, let's look at ethylene, C_2H_4. Looking at **Figure 8.3**, we can see the Lewis structure on the left (a); the central carbons are sp^2 hybridized. That means that the 2s orbital has hybridized with two of the three 2p orbitals. We can see the sp^2 hybrid orbitals on the second structure, (b). Two of the sp^2 orbitals overlap to form the σ carbon–carbon single bond. The two remaining unhybridized p orbitals are shown in the third structure, (c). They side overlap to form a π bond, as shown in the right structure (d). The combination of the σ and π bonds results in the carbon–carbon double bond.

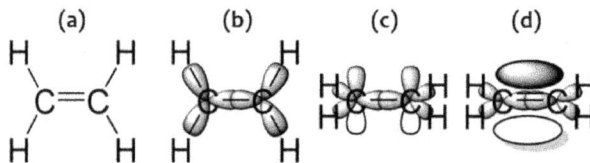

FIGURE 8.3 Double-bond hybridization for ethylene.

4. In a triple bond, how many σ and π bonds would you have?

5. How many p orbitals would be unhybridized on the central atom(s)?

6. What would be the EPG of the central atom(s)?

Molecular orbital (MO) theory takes a different approach to bonding theory. It states that a new molecular orbital forms for every atomic orbital. Unlike valence bond theory, MO theory states that the newly formed molecular orbitals are delocalized across the entire molecule. Whereas valence bond theory is very visual with illustrations of bonding, MO theory relies on diagrams, called MO diagrams. **Figure 8.4** is the MO diagram for dilithium, Li_2. On the far left and right of the diagram are the atomic orbitals (1s and 2s) contributed by each lithium atom. In the center, the new molecular orbitals (σ_{1s}, σ^*_{1s}, etc.) are formed as the two lithium atoms bond. The key to MO theory is the number of molecular orbitals must equal the number of atomic orbitals.

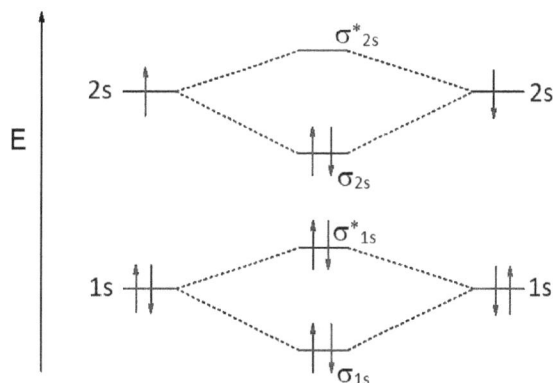

FIGURE 8.4 MO diagram for dilithium.

When two atomic orbitals come together, there is a probability that they will form a bond and a probability that they will not form a bond. The molecular orbitals that form are called *bonding molecular orbitals*, represented by σ or π. If the two atomic orbitals do not bond, they are called *antibonding molecular orbitals*, represented by σ* or π*. Bonding orbitals are always lower in energy than the atomic orbitals. Remember! Bond formation is *exothermic*, so a release in energy forms a lower energy molecular orbital than the corresponding atomic orbital. Antibonding orbitals are higher in energy than their atomic orbital counterparts.

Two other important aspects of MO diagrams are the identification of the *highest occupied molecular orbital* (HOMO) and *lowest unoccupied molecular orbital* (LUMO). The HOMO is the *highest* energy molecular orbital that *does* contain electrons, and the LUMO is the *lowest* energy molecular orbital that *does not* contain electrons. In the MO diagram of dilithium, the HOMO is σ_{2s} and the LUMO is σ^*_{2s}. The LUMO is always the next molecular orbital above the HOMO.

Use the following formula to calculate the bond order in an MO diagram. It is basically one half the bonding electrons (electrons in bonding molecular orbitals) minus the antibonding electrons (electrons in antibonding molecular orbitals).

$$\text{Bond order} = \frac{1}{2}\left(\text{bonding electrons} - \text{antibonding electrons}\right)$$

7. Explain why antibonding molecular orbitals are higher in energy than their atomic counterparts.

Review Questions

8. List the similarities and differences between valence bond theory and MO theory.

9. Why do we need hybridization in valence bond theory?

Application Questions

10. Count the number of σ and π bonds in the structure of *trans*-cinnamic acid and list the hybridization of the highlighted atoms.

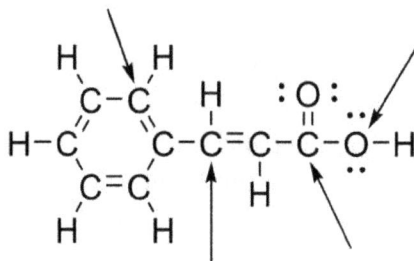

11. Which of the following molecules contain π bonds? (Circle all that apply.)

    NH_3 CH_2O CO_3^{2-} C_6H_6

12. What orbitals of phosphorous would be hybridized to form PF_5?

13. Chlorine and oxygen form a series of ions: ClO^- (hypochlorite), ClO_2^- (chlorite), ClO_3^- (chlorate), and ClO_4^- (perchlorate). These ions are used in many applications in everyday life, such as bleach. Draw the Lewis structures of the ions, and then determine the EPG and the hybridization of each chlorine atom.

14. Does the hybridization of each chlorine change as more oxygens are added in the series of ions? Why or why not?

15. Identify the HOMO and LUMO of diboron based on the MO diagram.

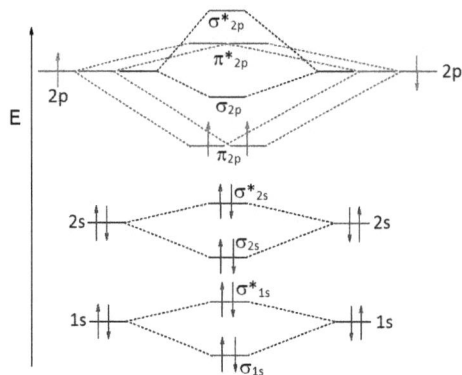

16. Calculate the bond order of diboron based on the MO diagram from question 15.

Figure Credits

CHAPTER 9

Molecular Geometry

Why do oil and water not mix?

Why do some molecules benefit humans, while others are harmful?

Learning Outcomes

- Identify the molecular geometry/structure and bond angles of various molecules and central atoms using valence shell electron pair repulsion (VSEPR) theory.
- Determine the hybridization associated with various molecular geometries/structures.
- Use bond polarities to explain a dipole moment and molecular polarity.
- Identify molecules that possess dipole moments.

Prerequisite Knowledge

- Chapter 7: Lewis Dot Structures
- Chapter 8: Bonding Theory

Background Information and Questions

Molecular geometry, sometimes referred to as molecular structure, is the classification of a molecule's three-dimensional shape based on the number of electron-rich regions around a central atom. Regions of electron density, also known as electron-rich regions or electron domains, can be bonds (single, double, or triple) as well as lone pairs. These regions arrange themselves around a central atom according to the valence shell electron pair repulsion (VSEPR) theory.

VSEPR theory states that electron pairs will arrange themselves around an atom to minimize the repulsions between themselves. Because these electron pairs can arrange themselves around spherical atoms, Lewis dot structures are not accurate representations of three-dimensional molecules because Lewis dot structures are two-dimensional representations. **Figure 9.1** illustrates how a Lewis dot structure is misleading in its representation of what a molecule actually looks like.

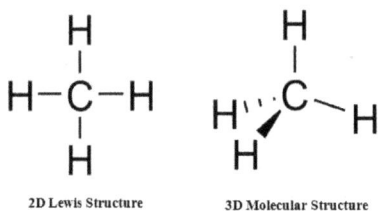

2D Lewis Structure   3D Molecular Structure

FIGURE 9.1 Lewis structure versus molecular structure.

The three-dimensional molecular structure in Figure 9.1 introduces two new bond representations. They are called the *wedge bond* (➤) and the *dash bond* (⠿). The wedge bond would be coming out of the plane of the paper, whereas the dash bond would be going behind the plane of the paper. In other words, the wedge bond is pointing out "toward you," whereas the dash bond is pointing "away from you" and into whatever surface the structure is drawn.

The two-dimensional Lewis structure implies that the bond angle between the H–C–H atoms is 90°, but it is misleading. The three-dimensional molecular structure clearly indicates that the bond angle is larger than 90°; in fact, it is actually 109.5°. That is the farthest apart that four domains, or bonds in this case, can be arranged around a sphere or central atom. Domains can be single bonds, multiple bonds, or lone pairs.

When identifying the three-dimensional structure of a molecule, there are two classifications: electron pair geometry (EPG) and molecular geometry (MG), also called the molecular structure. EPG is the total number of electron domains (bonds and lone pairs). MG is focused on the number of bonded domains (bonds). **Table 9.1** illustrates the various domains and arrangements around the central atom. It also identifies the EPG and MG, relates domains to EPG and MG, and shows bond angles for EPGs. Remember that electron-dense regions want to be as far apart from each other as possible, so these are the largest bond angles possible for the number of domains.

Note that lone pair domains reduce the bond angles. CH_4, NH_3, and H_2O all have a tetrahedral EPG but have very different molecular geometries: tetrahedral, trigonal planar, and bent, respectively. The bond angles are 109.5° for tetrahedral, 107.5° for trigonal pyramidal, and 104.5° for bent. The lone pair domains reduce the angle between the bonding domains because of their increased size.

TABLE 9.1 Three-Dimensional Structure of Molecules

	Number of Total Domains	Electron Pair Geometry	Number of Bonded Domains	Number of Lone Pair Domains	Molecular Geometry
	2	Linear	2	0	Linear
	3	Trigonal planar	3	0	Trigonal planar
			2	1	Bent
	4	Tetrahedral	4	0	Tetrahedral
			3	1	Trigonal pyramidal
			2	2	Bent
	5	Trigonal bipyramidal	5	0	Trigonal bipyramidal
			4	1	Seesaw

(continued)

TABLE 9.1 Three-Dimensional Structure of Molecules (*Continued*)

	Number of Total Domains	Electron Pair Geometry	Number of Bonded Domains	Number of Lone Pair Domains	Molecular Geometry
			3	2	T-shaped
			2	3	Linear
	6	Octahedral	6	0	Octahedral
			5	1	Square pyramidal
			4	2	Square planar
			3	3	T-shaped
			2	4	Linear

To determine the EPG and MG:

I. Draw the Lewis dot structure of the molecule.

II. Count the total number of electron domains (lone pairs *and* bonds) around the central atom (remember that double and triple bonds count as one domain), and use the information in Table 9.1 to assign the EPG.

III. Determine how many of the electron domains are nonbonding (lone pairs), and assign the molecular geometry from Table 9.1.

Notice that there are no bond angles for the trigonal bipyramidal electron pair geometry structure. This is because the same bond angle does not separate all five bonds or domains. Assign bond angles to the trigonal bipyramidal structure that has five bonded domains. There are two different angles you will need to determine.

1. Determine the EPG and the MG of a molecule that has two bonded domains and two lone-pair domains.

2. What is the molecular geometry of NH_3? Are the MG and the EPG the same for NH_3?

3. Why would NH_4^+ have a bond angle of 109.5°, given that NH_3 has a bond angle of 107.5°?

Molecular polarity is based on both bond polarity and molecular geometry. We will be using bond polarity and molecular geometry to determine whether an entire molecule is polar or has a dipole moment.

Recall that bond polarity represents electrons being unequally shared in a covalent bond. This is the result of a large difference (> 1.0) in the electronegativities of the atoms in the bond (Chapter 6). To determine the polarity of an entire molecule, both bond polarities and dipole moment directions matter. Now, here is a tricky statement: "In order for a molecule to be polar, it must have polar bonds; however, all molecules with polar bonds may not be polar." This is due to the molecular geometry of the molecule.

To determine the molecular polarity, we will need to add the bond dipoles, like vectors. If you are not sure about vectors, just think about the sum of the overall pulls of the electronegative atoms. Looking at the molecule shown in **Figure 9.2**, we can determine if it is polar by considering the bond polarity of the only bond.

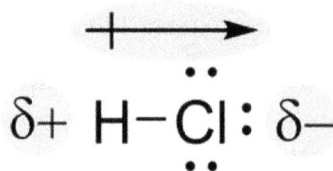

FIGURE 9.2 Determination of molecular polarity of hydrogen chloride.

When determining molecular polarity, one must consider the three-dimensional shape of the molecule. The summation of the bond dipoles will determine whether the entire molecule is polar. Consider the molecule in **Figure 9.3**. The two bond dipoles (left) sum to the overall dipole moment (right) for CH_2Cl_2.

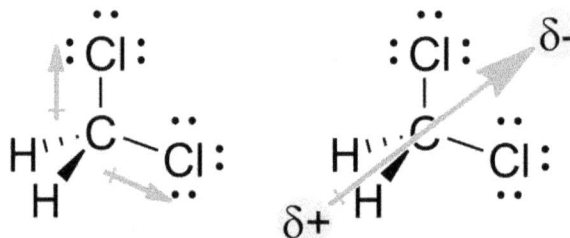

FIGURE 9.3 Determination of molecular polarity of dichloromethane.

Polarity helps us determine the properties of a molecule. Will it be water soluble? Water is polar and interacts best with other polar molecules. Compounds such as olive oil are not water soluble because they do not have a dipole moment (i.e., they are nonpolar).

4. What does a dipole moment in a molecule represent?

5. Draw the dipole moment for a PCl_3 molecule. Do you think the dipole moment will be as large as that for the PF_3 molecule? Why or why not?

Review Questions

6. Why do molecules have certain shapes?

7. How does molecular geometry (MG) contribute to a molecule's dipole moment or polarity?

Application Questions

8. Compare and contrast the EPG and MG of SF_4 and SF_6. Are they the same or different? Which one would be polar? Which one would be nonpolar?

9. Rationalize (explain) the trend in dipole strengths (μ): for CH_3Cl, $\mu = 1.92$ D; for CH_2Cl_2, $\mu = 1.60$ D; for $CHCl_3$, $\mu = 1.04$ D; and for CCl_4, $\mu = 0.00$ D.

10. Complete the following chart.

Formula	Lewis Structure	3D Sketch	EPG	MG	Bond Angle	Polar or Nonpolar?
NH_3						
XeF_2						
SO_4^{2-}						
AsF_5						

11. *Trans*-cinnamic acid is derived from cinnamon; it is a biologically relevant molecule, and certain cinnamic acids can be powerful antioxidants. Determine the EPG, MG, and bond angle around the indicated atoms in *trans*-cinnamic acid.

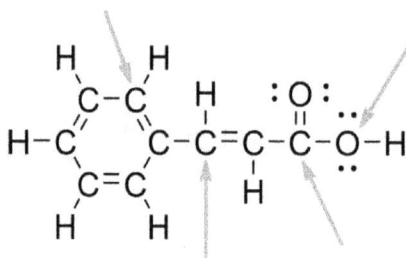

12. An unknown organic molecule has a molar mass of 74.12 g/mol and was found to contain 64.81% C, 13.60% H, and 21.59% O. Determine the bond angle between the bonded regions around the oxygen atom(s).

Reflection Questions

13. Why might structure be important when considering the role of molecules such as oxygen and carbon monoxide in aiding and harming the human body?

14. Water is polar and can dissolve or interact with other polar molecules. Oil is nonpolar and interacts with other nonpolar molecules. If this is true, then how do you clean grease or oil with water? *Hint! Is there something you have to add to help clean the grease or oil?*

Figure Credits

CHAPTER 10

The Periodic Table

What information can we garner from the periodic table?

How does the arrangement of atoms on the periodic table relate to electron configurations?

Learning Outcomes

- Identify an element's period and group using the periodic table.
- Determine an element's chemical and physical properties based on its location on the periodic table.
- Rationalize periodic trends using effective nuclear charge and electron configurations.
- Predict atomic radius, first ionization energy, and first electron affinity of elements based on periodic trends.

Prerequisite Knowledge

- Chapter 3: Atomic Structure
- Chapter 5: Electronic Structure II

Background Information and Questions

Dmitri Mendeleev is credited as the father of the periodic table; however, Lothar Meyer also had a similar discovery at the same time. There were other versions of the periodic table prior to their unique way of arranging the elements. However, Mendeleev and Meyer were the first to recognize that elements could be arranged by increasing atomic weight.[1] Both Mendeleev and Meyer noted that chemical and physical properties repeat periodically when elements are arranged this way. Keep in mind that atomic number had not yet been discovered.

The periodic table is labeled uniquely. Columns of elements are called *groups* or *families*; rows of elements are called *periods*. See the periodic table in **Figure 10.1**. Periods are labeled by integers (1, 2, 3, …); groups may be labeled by integers (1–18) or by Roman numerals (**Figure 10.2**). The taller groups (groups 1, 2, and 13–18) are known as *main groups*.

> You may want to have a periodic table available for viewing as you read this chapter. The following periodic tables are not complete and only illustrate specific features being discussed.

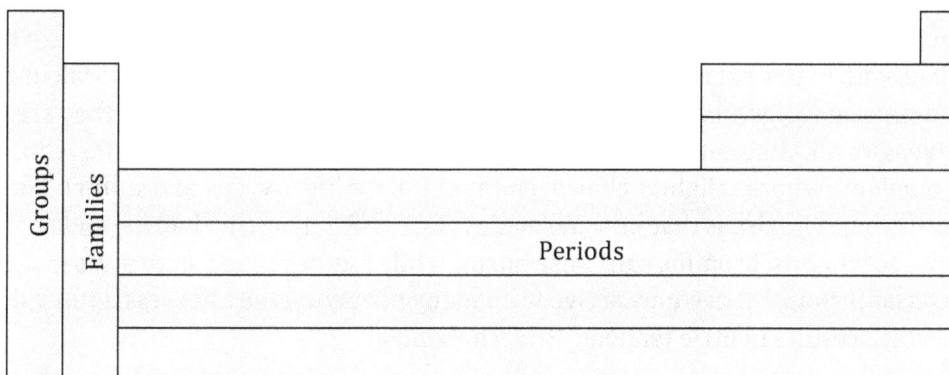

FIGURE 10.1 Periodic table groups, families, and periods.

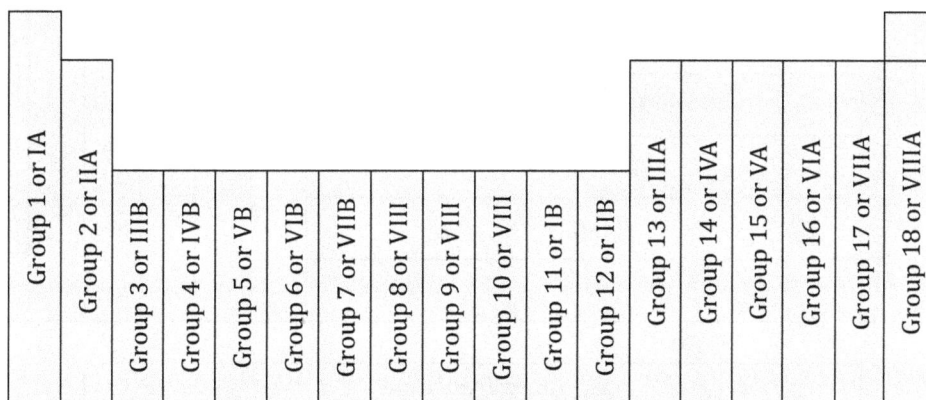

FIGURE 10.2 Numbering of periodic table groups.

1 Orna, M.V. Julius Lothar Meyer and Dmitri Ivanovich Mendeleev. Science History Institute. https://www.sciencehistory.org/historical-profile/julius-lothar-meyer-and-dmitri-ivanovich-mendeleev

1. What is the period and group number of the following elements?

    Se   Co   Xe   Ra

Elements are arranged by increasing atomic number, which gives rise to trends in chemical and physical properties in the groups. Some groups have been given names due to their unique properties. You can follow along with the names of these groups on the periodic table in **Figure 10.3**. I will discuss the groups from left to right; as each group is introduced, some health benefits and drawbacks will be briefly discussed as well.

Group 1 elements are known as the *alkali metals*. They all occur in nature and react violently with water to form oxides and, subsequently, hydroxides. Alkali metals become more reactive as you move down the group. Sodium in the ionic form is necessary for life. Group 2 elements are called *alkaline earth metals*; these metals also react to form basic solutions with water but not violently like the alkali metals. Magnesium and calcium are essential elements for human health. Group 16 elements are the *chalcogens*; they are very reactive with alkaline earth metals. Heavier chalcogens are toxic (selenium, tellurium, and polonium), whereas lighter chalcogens are vital for life (oxygen and sulfur). The *halogens* in group 17 exist as diatomic molecules (F_2, Cl_2, Br_2, I_2, etc.). Fluorine and chlorine are toxic to humans; bromine can cause burns, while iodine is used as an antiseptic. The *noble gases* (group 18) are unreactive with many elements; they possess a full valence shell, which results in little tendency to form bonds.

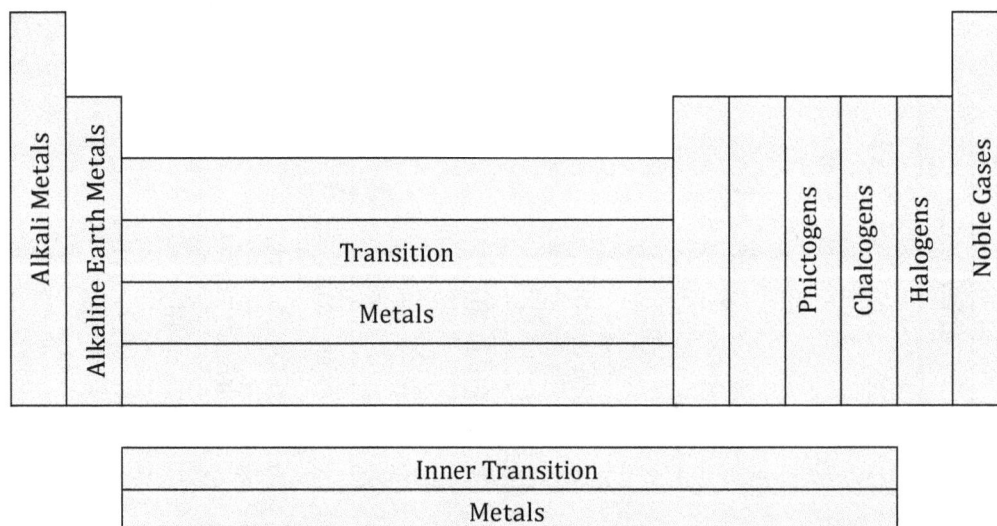

FIGURE 10.3 Groups named based on their unique properties.

2. List two alkali metals and two chalcogens.

Let's examine the physical properties of metals, nonmetals, and metalloids. You can see the areas of these elements on the periodic table in **Figure 10.4**. Metals are on the left side of the periodic table; nonmetals are on the right side. Metalloids lay on the border between these areas. The same trend is observed for their physical properties as well. Metals, except for mercury, are all solids at room temperature; most are a shiny silver color and hard. Metals are good conductors of heat and electricity; they are ductile (can easily be drawn into wires) and malleable (can be hammered into thin sheets). Many nonmetals are gases at room temperature; they are dull and brittle. Nonmetals are insulators; they are poor conductors of heat and electricity. Metalloids have physical properties that are a mixture of metals and nonmetals; for example, metalloids, like silicon, are semiconductors of electricity.

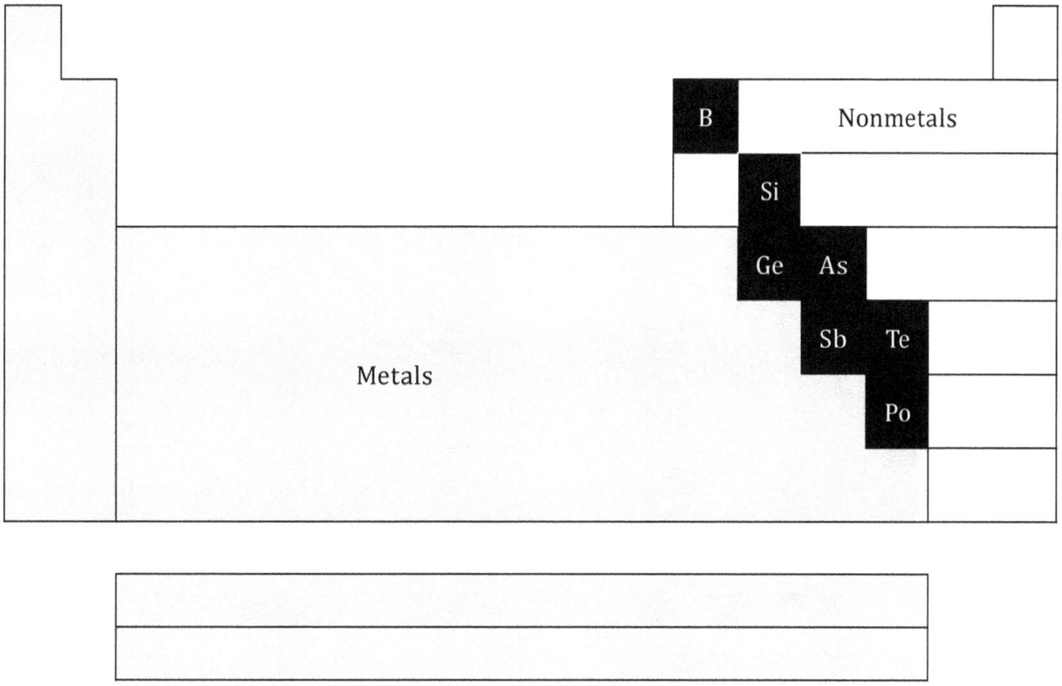

FIGURE 10.4 Metals, nonmetals, and metalloids.

3. List and describe the characteristics of metals.

By examining the arrangement of elements, we can note some interesting trends. Let's discuss the trends of atomic radius, first ionization energy, and electron affinity. Please keep in mind that these are generalized trends; there are exceptions to these trends.

Atomic radius is the distance from the center of the nucleus to the edge of an atom's electron cloud; effectively, it tells us the size of an atom. Atomic radius increases as you move down a group and increases as you move to the left across a period (see **Figure 10.5**). The atomic radius periodic trend stems from two sources: addition of electrons to higher energy levels as you move down a period and an increase in the *effective nuclear charge* as you move from left to right across a period. As one moves down a group, electrons are added to higher energy levels that are farther from the nucleus. This results in a larger atomic radius.

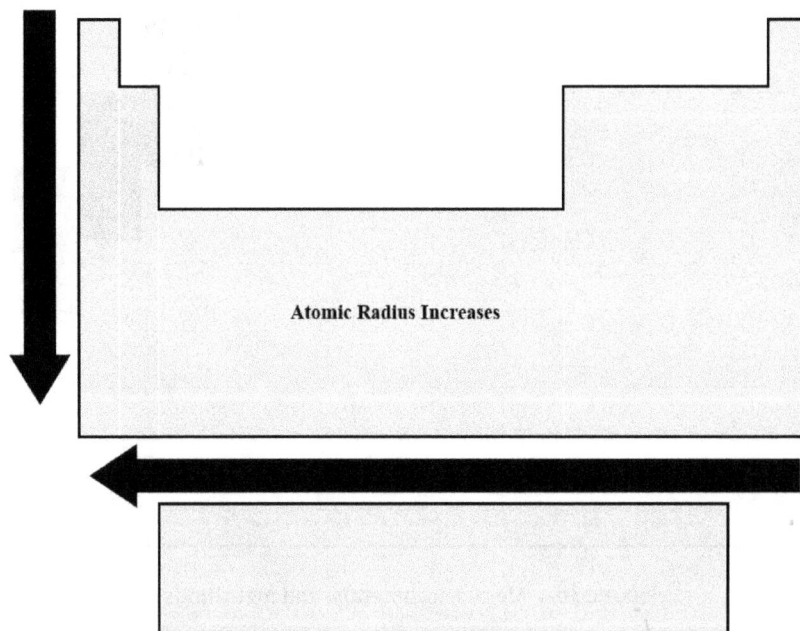

Atomic Radius Increases

FIGURE 10.5 Periodic table trends in atomic radius.

To explain the left-to-right trend of atomic radius, we use effective nuclear charge, which is the net positive charge experienced by a specific electron in an atom. The equation for the effective nuclear charge is:

$$Z_{eff} = Z - S$$

Z_{eff} is the symbol for effective nuclear charge, Z is the atomic number, and S represents shielding. Shielding is the number of electrons between the electron in question and the nucleus. As you move from left to right across the periodic table, the effective nuclear charge increases because the number of protons (Z) are increasing, while electrons are added to the same energy level. This results in no change to the S value. The increase in Z_{eff} results in a smaller atomic radius.

4. What would happen to the effective nuclear charge as the number of electrons between the electron in question and the nucleus increases?

5. Why would Se have a smaller radius than Sc?

The *first ionization energy* is the amount of energy required to remove 1 mole of electrons from 1 mole of gaseous atoms, which is illustrated by the following equation:

$$X (g) \rightarrow X^+ (g) + e^-$$

First ionization energy increases as you move to the right across a period and up a group or family (**Figure 10.6**). The trend can be rationalized by thinking about some of the same factors used to consider the atomic radius trend. The magnitude of the nucleus's positive charge, the distance the electron is from that nucleus, and the number of electrons between it and the nucleus are all factors that affect ionization energy. The ionization energy trend is the opposite of the atomic radius trend. Ionization requires energy to remove electrons; larger atoms require less energy to remove the electron because the outer electron is farther from the nucleus and experiences a lower effective nuclear charge. The opposite is true for smaller atoms; they have higher first ionization energies because the electrons experience more of the nucleus's positive charge. Chemists also measure and rationalize second, third, et cetera, ionization energies.

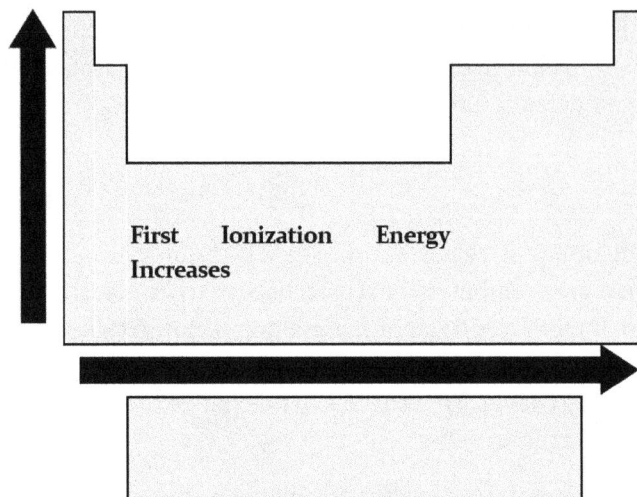

FIGURE 10.6 Periodic table trends in first ionization energy.

6. Why would F have a greater ionization energy than Br?

First electron affinity is the energy released when one mole of electrons is added to one mole of gaseous atoms:

$$X\,(g) + e^- \rightarrow X^-\,(g)$$

First electron affinity follows the same trend direction as ionization energy (**Figure 10.7**). It is a measure of the affinity of a nucleus for an electron. Smaller atoms will have nuclei that have a stronger affinity for an electron. This is because there is a higher effective nuclear charge that could be experienced by the incoming electron. Therefore, smaller atoms have higher electrons affinities (release more energy upon addition of an electron) than larger atoms.

7. Why would cesium have a smaller electron affinity than chlorine?

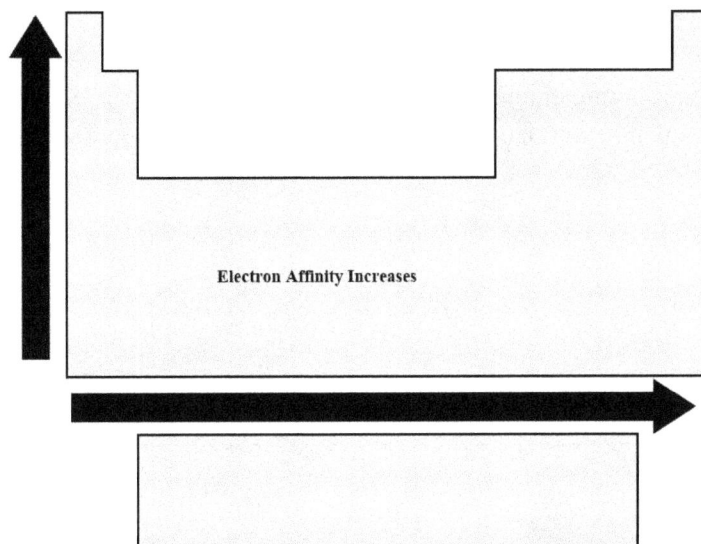

Electron Affinity Increases

FIGURE 10.7 Periodic table trends in first electron affinity.

Review Questions

8. Why do the members of each group or family have similar chemical and physical properties?

9. How does effective nuclear charge contribute to periodic trends?

10. How would ionization affect the radius of an atom? Discuss why the radius would change with the formation of a cation and an anion.

Application Questions

11. For the following element boxes labeled i–iv, identify the element; label each as a metal, nonmetal, or metalloid; and write the group name (if there is one) or number.

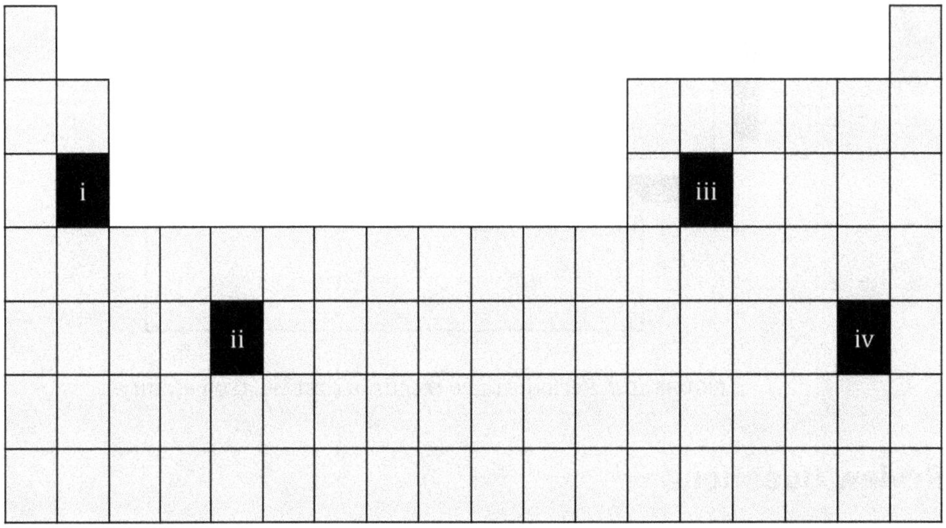

12. Calculate the effective nuclear charge (Z_{eff}) for carbon.

13. Place the following atoms in order of *decreasing* atomic radius.

    (a) Ge, Cl, Br

    (b) P, N, O

    (c) Rb, Ca, Fr

14. Write a chemical equation representing the second ionization energy of magnesium. Would you expect the process to release energy or to require energy?

15. Determine which of the following atoms would have the highest first ionization energy.

    (a) S, Co, Ca, K

    (b) As, Zn, O, Hg

    (c) Cs, Mn, Si, F

16. Which element would you expect to exhibit the higher electron affinity?

    (a) H or He

    (b) Al or Si

    (c) Ca or Sr

CHAPTER 11

Chemical Nomenclature

How do we give chemicals their names?

How can a name tell us the properties of a chemical?

Learning Outcomes

- ▪ Recognize common polyatomic ions by name and formula.
- ▪ Name ionic compounds from formulas and derive formulas from names.
- ▪ Explain why transition metal names require Roman numerals.
- ▪ Name molecular compounds from formulas and derive formulas from names.
- ▪ Name acids and bases from formulas and derive formulas from names.

Prerequisite Knowledge

- ▪ Chapter 6: Chemical Bonding

Background Information and Questions

Nomenclature can be quite confusing. Especially since it is presented all at once for at least four classes of chemicals. In this lesson, we will discuss how to name ionic compounds, acids, bases, and molecular compounds. Before naming these compounds, you will need to learn to recognize them. Ionic compounds are usually composed of a metal and a nonmetal. Acids typically have hydrogen as the first element in their formula; bases usually feature OH^- or hydroxide in their formulas. Molecular or covalent compounds are composed only of nonmetals.

1. How does one identify acids, bases, and ionic and molecular compounds?

Before we learn to name chemicals, we need to learn to write the correct formula. Ionic compounds are made up of ions. Ions have a different number of electrons than protons. Cations have lost electrons when compared to their neutral-atom counterparts. Anions have gained electrons when compared to their neutral-atom counterparts. Note how the radius changes upon the gain or loss of an electron (**Figure 11.1**).

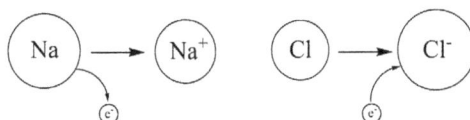

FIGURE 11.1 Change in atomic radius with gain or loss of electron.

When ions combine to form an ionic compound, they are held together by the electrostatic forces between the cations and anions. The compounds always form a neutral compound. The positive cations and negative anions balance each other out, neutralizing the charges. An ionic compound is always neutral. If we make a compound from Mg^{2+} and F^-, then we will need two fluoride anions to neutralize the 2+ charge on the magnesium ion; therefore, the formula will be MgF_2. One does not have to think about balancing charges every time; there is a shortcut called the *crisscross method.* One simply crisscrosses the charges and uses those numbers and the subscripts for the other ion (**Figure 11.2**).

$$Na^+ + O^{2-} \longrightarrow Na^{\oplus}_2O^{2\ominus} \longrightarrow Na_2O$$

FIGURE 11.2 Crisscross method.

The rules for writing ionic formulas are as follows:

I. The cation is always written first, followed by the anion.

II. In all ionic compounds, a subscript of one (1) is understood and never written.

III. All ionic compounds are reduced by their least common denominator. If both ions have the same charge, then divide the subscripts by themselves and reduce the subscripts to one. For example, $Al^{3+} + N^{3-} \rightarrow Al_3N_3 \rightarrow AlN$.

IV. If you have a polyatomic (many atoms) ion, treat it as if were one atom. Do NOT change its subscripts when writing an ionic compound.

V. If the formula requires a polyatomic ion to have a subscript, put the polyatomic in parentheses, like $Ca(ClO_2)_2$. This assumes there are two entire chlorite anions for every calcium ion. Otherwise, it would appear like $CaClO_{22}$. There are NOT 22 oxygens in the formula.

VI. You only need parentheses if a polyatomic ion has a subscript; the following compounds do not require parentheses because the polyatomic ions do not have a subscript: $Na^+ + SO_4^{2-} \rightarrow Na_2SO_4$ and $Mg^{2+} + SO_3^{2-} \rightarrow MgSO_3$.

VII. In order to determine the ions that make up an ionic compound, simply reverse the crisscross method to determine the charge on each of the ions.

Use the periodic table of common ionic charges in **Figure 11.3** and the table of common polyatomic ions (**Table 11.1**) to answer the following questions, as well as the rest of the questions in this chapter.

2. In the name of an ionic compound, which ion is always written first, the *anion* or the *cation*?

3. Using the periodic table in Figure 11.3, which types of elements tend to form cations, and which types of elements tend to form anions?

1	2	3	4	5	6	7	8	9	10	11	12	13	14	15	16	17	18
hydrogen 1 H^+																	helium 2 He
lithium 3 Li^+	beryllium 4 Be^{2+}											boron 5 B	carbon 6 C	nitrogen 7 N^{3-}	oxygen 8 O^{2-}	fluorine 9 F^-	neon 10 Ne
sodium 11 Na^+	magnesium 12 Mg^{2+}											aluminium 13 Al^{3+}	silicon 14 Si	phosphorus 15 P^{3-}	sulphur 16 S^{2-}	chlorine 17 Cl^-	argon 18 Ar
potassium 19 K^+	calcium 20 Ca^{2+}	scandium 21 Sc^{3+}	titanium 22 Ti 4+, 3+, or 2+	vanadium 23 V 5+,4+,3+,2+	chromium 24 Cr 6+, 3+, or 2+	manganese 25 Mn^{7+} 6+,4+,3+,2+	Fe^{3+} Fe^{2+}	Co^{2+} Co^{3+}	Ni^{3+} Ni^{2+}	Cu^{2+} Cu^+	zinc 30 Zn^{2+}	gallium 31 Ga^{3+}	germanium 32 Ge	arsenic 33 As 5+, 3+,3-	selenium 34 Se^{2-}	bromine 35 Br^-	krypton 36 Kr
rubidium 37 Rb^+	strontium 38 Sr^{2+}	yttrium 39 Y^{3+}	zirconium 40 Zr^{4+}	vanadium 41 Nb^{5+} Nb^{3+}	molybdenum 42 Mo^{6+}	technetium 43 Tc 7+, 6+, or 4+	Ru^{4+} Ru^{3+}	rhodium 45 Rh^{3+}	Pd^{4+} Pd^{2+}	silver 47 Ag^+	cadmium 48 Cd^{2+}	indium 49 In^{3+}	Sn^{4+} Sn^{2+}	Sb^{5+} Sb^{3+}	Tellurium 52 Te^{2-}	iodine 53 I	xenon 54 Xe
caesium 55 Cs^+	barium 56 Ba^{2+}	lanthanum 57 La^{3+}	hafnium 72 Hf^{4+}	tantalum 73 Ta^{5+}	tungsten 74 W^{6+}	rhenium 75 Re 7+, 6+, or 4+	Os^{4+} Os^{3+}	Ir^{4+} Ir^{3+}	Pt^{4+} Pt^{2+}	Au^{3+} Au^+	Hg^{2+} Hg_2^{2+}	Tl^{3+} Tl^+	Pb^{4+} Pb^{2+}	Bi^{5+} Bi^{3+}	polonium 84 Po 4+, 2+, or 3-	astatine 85 At^-	radon 86 Rn
francium 87 Fr^+	radium 88 Ra^{2+}	actinium 89 Ac^{3+}															

FIGURE 11.3 Common ion charges.

TABLE 11.1 Common Polyatomic Ions

acetate	$C_2H_3O_2{}^-$ / CH_3COO^-	hydrogen carbonate (bicarbonate)	$HCO_3{}^-$	hydrogen sulfite (bisulfite)	$HSO_3{}^-$	permanganate	$MnO_4{}^-$	hydroxide	OH^-
cyanide	CN^-	carbonate	$CO_3{}^{2-}$	hydrogen sulfate (bisulfate)	$HSO_4{}^-$	cyanate	OCN^-	oxalate	$C_2O_4{}^{2-}$
hypochlorite	ClO^-	dihydrogen phosphite	$H_2PO_3{}^-$	sulfite	$SO_3{}^{2-}$	thiocyanate	SCN^-	chromate	$CrO_4{}^{2-}$
chlorite	$ClO_2{}^-$	hydrogen phosphate	$HPO_3{}^{2-}$	sulfate	$SO_4{}^{2-}$	peroxide	$O_2{}^{2-}$	dichromate	$Cr_2O_7{}^{2-}$
chlorate	$ClO_3{}^-$	phosphate	$PO_4{}^{3-}$	thiosulfate	$S_2O_3{}^{2-}$	nitrite	$NO_2{}^-$	amide	$NH_2{}^-$
perchlorate	$ClO_4{}^-$	phosphite	$PO_3{}^{3-}$	silicate	$SiO_3{}^{2-}$	nitrate	$NO_3{}^-$	ammonium	$NH_4{}^+$

Now that you've learned how to write compounds, it is time to learn how to name them. Use the charts in **Figures 11.4** to **11.6** to name different compounds.

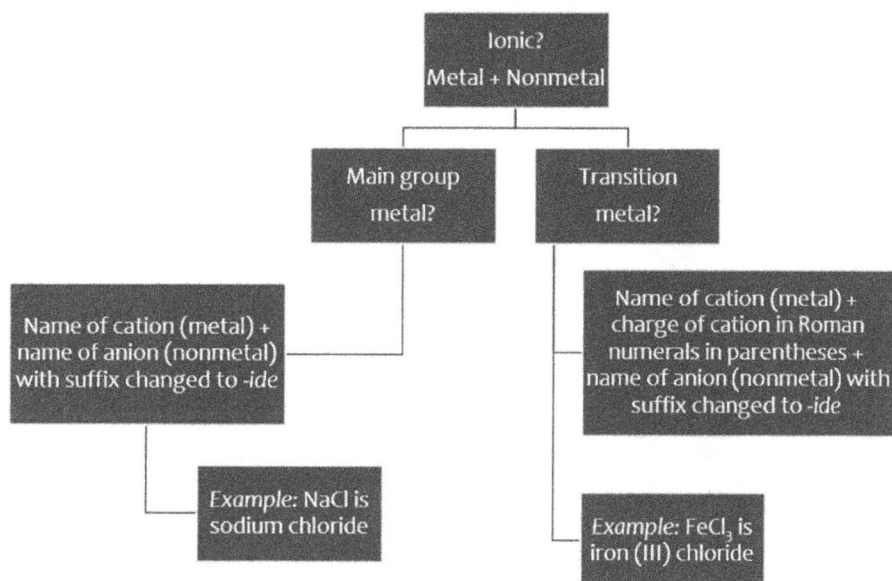

FIGURE 11.4 Naming of ionic compounds.

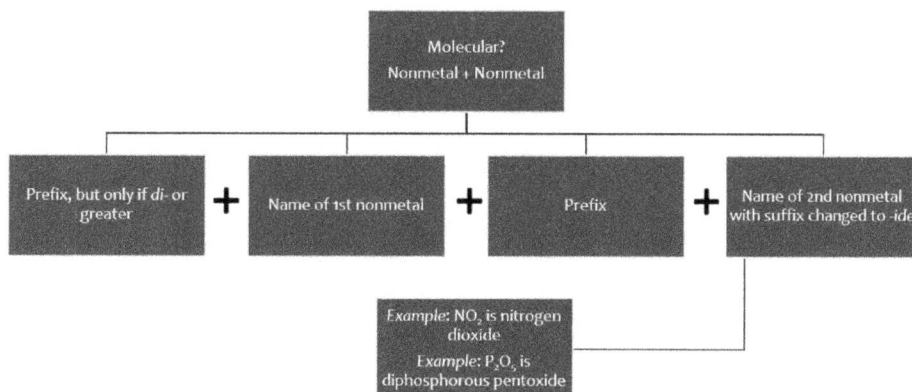

FIGURE 11.5 Naming of molecular compounds.

Molecular compounds do not have charges, so there is no need for a discussion about how to write their formulas. This is why their names require prefixes, which tell us how many of each atom is in the formula. See **Table 11.2** for a list of prefixes.

TABLE 11.2 Prefixes for Molecular Compounds

1	2	3	4	5	6	7	8	9	10
mono-	di-	tri-	tetra-	penta-	hexa-	hepta-	octa-	nona-	deca-

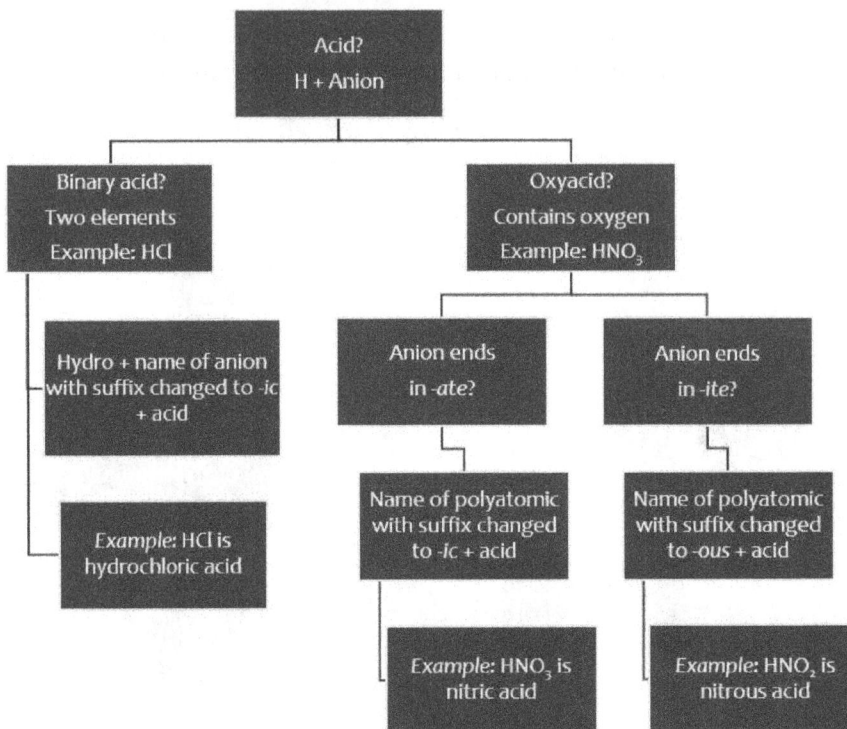

FIGURE 11.6 Naming of acids.

Acids are molecular compounds because they do not contain a metal; however, their formulas can be written using the *crisscross method*, like ionic compounds.

4. For nonmetals, how does the name of the anion differ from the name of the element?

5. Using the periodic table of common ionic charges, why might transition metals require a Roman numeral in their name?

6. What prefix would be used to indicate that a molecular compound has seven of a particular atom?

Review Questions

7. When they form from neutral atoms, do cations gain or lose electrons?

8. When they form from neutral atoms, do anions gain or lose electrons?

9. How many electrons would rubidium gain or lose when forming an ion?

10. Once rubidium gained or lost its electrons, which atom would have the same number of electrons?

11. Consider the name calcium bromide, $CaBr_2$.

    (a) What is the charge on the calcium ion?

    (b) What is the charge on each bromide ion?

    (c) How many bromide ions are in the formula?

    (d) What is the overall (total) charge on the compound?

Application Questions

12. Classify each of the following compounds as an ionic compound, an acid, a base, or a molecular compound.

    (a) NaCl

    (b) CO_2

    (c) HCl

    (d) KOH

    (e) SO_3

    (f) NH_4NO_3

13. Write the formulas for the ionic compounds formed from the following ions:

    (a) magnesium and sulfate

    (b) calcium and fluoride

    (c) sodium and phosphide

    (d) sodium and phosphite

(e)   sodium and phosphate

(f)   ammonium and sulfate

14.   Which ions make up the following ionic compounds?

(a)   NaCl

(b) $Ca_3(PO_4)_2$

(c) KNO_3

(d) MnS_2

(e) Fe_2O_3

(f)   FeO

15.   Name and identify each of the following as an acid, a base, an ionic compound, or a molecular compound.

(a) H_2SO_4

(b) PbO_2

(c) SO_2

(d) KOH

(e) CsCl

(f) H_2S

16. What is wrong with the following compound names?

   (a) mononitrogen dioxide, NO_2

   (b) magnesium dichloride, $MgCl_2$

   (c) sulfuric acid, H_2SO_3

   (d) iron oxide, FeO

   (e) sodium nitrate, Na_3N

   (f) dinitrogen pentoxide, N_2O_4

CHAPTER 12

Chemical Reactions I

How do we write and identify chemical reactions?

What types of reactions are the result of specific types of chemicals?

Learning Outcomes

▣ Balance chemical equations.
▣ Predict the products of chemical reactants.

Prerequisite Knowledge

▣ Chapter 10: The Periodic Table
▣ Chapter 11: Chemical Nomenclature

Background Information and Questions

Chemical reactions are keeping you alive; they help maintain homeostasis in all living organisms! They are one of the most important aspects of the study of chemistry. We represent chemical reactions using symbolic notation. Chemicals that exist prior to the reaction are called *reactants*; chemicals that are made during (and exist after) the reaction are called *products*. See **Figure 12.1**.

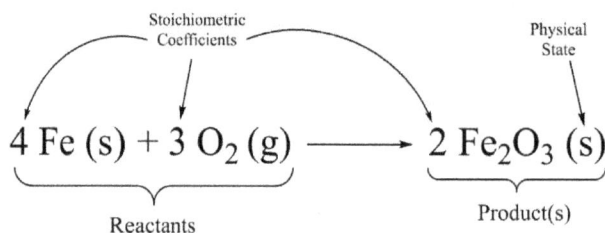

FIGURE 12.1 Representation of a chemical reaction.

Referring to Figure 12.1, we see there are also *physical states*. Chemicals exist as solids (*s*), liquids (*l*), gases (*g*), or dissolved in water (*aq*). Because matter can neither be created nor destroyed, we must balance a chemical equation to make sure that there are an equal number of atoms on both sides of the reaction arrow. We use *stoichiometric coefficients* to balance the equation to ensure that the law of conservation of matter is upheld. These coefficients are numbers placed in front of the compound in the balanced chemical equation; they also represent the ratios between the chemicals in a chemical reaction.

1. What do the stoichiometric coefficients represent in a chemical reaction?

How does one balance a chemical equation? To answer that, let's look at the following equation:

$$H_2 \text{ (g)} + Cl_2 \text{ (g)} \rightarrow HCl \text{ (g)}$$

Note that there are two hydrogen and two chlorine atoms on the reactant side but only one hydrogen and one chlorine atom on the product side. In order to balance the equation, we need to add a coefficient of two (2) in front of the HCl (g) product. This will indicate two HCl molecules, which is two hydrogen atoms and two chlorine atoms. The balanced equation is thus as follows:

$$H_2 \text{ (g)} + Cl_2 \text{ (g)} \rightarrow 2 \text{ HCl (g)}$$

Another balancing technique is to list the atoms on both sides of the reaction arrow, as depicted in **Table 12.1**.

TABLE 12.1 Balancing an Equation by Listing the Atoms: Step 1

CCl_4 (l) + HF (aq) → CF_2Cl_2 (g) + HCl (aq)	
C = 1 Cl = 4 H = 1 F = 1	C = 1 Cl = 3(Total Cl = 2+1) H = 1 F = 2

To balance the equation in Table 12.1, let's look at the number of atoms on each side. One strategy is to start with the atom that has the highest subscript. In this case, chlorine has the highest subscript. The reactant side has four chlorine atoms and the product side has three (2 + 1). We can equalize that number by multiplying the one chlorine from HCl by 2, which is the same thing as adding a coefficient in front of the HCl (**Table 12.2**). Remember, this will also double the number of hydrogen atoms.

TABLE 12.2 Balancing an Equation by Listing the Atoms: Step 2

CCl_4 (l) + 2 HF (aq) → CF_2Cl_2 (g) + 2 HCl (aq)		
C = 1 Cl = 4 H = 1 F = 1	C = 1 Cl = 2 Cl = 1 × 2 = 2 H = 1 × 2 = 2 F = 2	2 +2 4 Cl

Now there are four chlorines on the reactant side and four chlorines (2 + 2) on the product side. However, our hydrogens are now unbalanced. Let's tackle that now. To balance the hydrogen atoms, let us add a coefficient of 2 in front of HF to balance the hydrogens (**Table 12.3**).

TABLE 12.3 Balancing an Equation by Listing the Atoms: Step 3

CCl_4 (l) + 2 HF (aq) → CF_2Cl_2 (g) + 2 HCl (aq)		
C = 1 Cl = 4 H = 1 × 2 = 2 F = 1 × 2 = 2	C = 1 Cl = 2 Cl = 1 × 2 = 2 H = 1 × 2 = 2 F = 2	2 +2 4 Cl

Now let's check to see if there are any other unbalanced atoms. Both sides have one carbon atom, four chlorine atoms, two hydrogen atoms, and two fluorine atoms; it seems the equation is balanced!

2. How many oxygen atoms are on both sides of the following equation? Is it balanced?

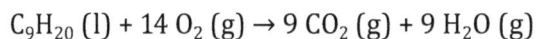

$$C_9H_{20} \text{ (l)} + 14\ O_2 \text{ (g)} \rightarrow 9\ CO_2 \text{ (g)} + 9\ H_2O \text{ (g)}$$

3. What would be the coefficient of Al if the following equation were balanced?

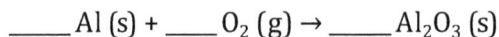

$$\underline{\hspace{1cm}} Al \text{ (s)} + \underline{\hspace{1cm}} O_2 \text{ (g)} \rightarrow \underline{\hspace{1cm}} Al_2O_3 \text{ (s)}$$

How does one predict what products a chemical reaction might produce if only given the reactants? There are several reaction predictors. They are listed in **Table 12.4** using generic letters in place of chemicals. Each solitary letter represents an element, and combinations of letters represent compounds. You can use the table to predict what products a reaction may produce.

TABLE 12.4 Predicting Products of Chemical Reactions

Reaction Predictor Name	General Predictor Equation
Synthesis/combination	A + B → AB
Decomposition/dissociation	AB → A + B
Single replacement	AB + C → AC + B
Double replacement	AB + CD → AD + CB

4. What type of substances (compounds or elements) are the products of decomposition reactions?

5. What type of substances (compounds or elements) are the products of synthesis reactions?

6. Provide an example of a balanced single replacement reaction. Do not copy the generic one above; use an example with balanced chemical formulas!

7. Provide an example of a balanced double replacement reaction. Do not copy the generic one above; use an example with balanced chemical formulas!

Review Questions

8. Why do we balance chemical equations?

9. Why do you think it is important to be able to predict the products of chemical reactions?

Application Questions

10. Balance the following equations:

   (a) $H_2 (g) + F_2 (g) \rightarrow HF$

(b) Cs (s) + H_2O (l) → CsOH (aq) + H_2 (g)

(c) $Mg(OH)_2$ (aq) + H_3PO_4 (aq) → $Mg_3(PO_4)_2$ (aq) + H_2O (l)

(d) Cu (s) + HNO_3 (aq) → $Cu(NO_3)_2$ + NO + H_2O

(e) C_6H_{14} (l) + O_2 (g) → CO_2 (g) + H_2O (g)

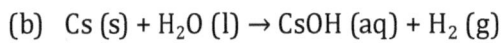

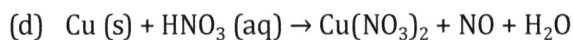

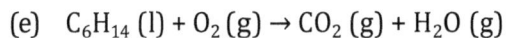

11. Predict the products from the following reactants.

(a) Ca (s) + Cl_2 (g) →

(b) Ag_2S →

(c) sulfuric acid and sodium hydroxide

(d) silver nitrate and potassium bromide

(e) solid zinc and copper (II) chloride

12. For each of the following, predict the products of the reactants and write a balanced chemical equation.

(a) lead (II) chloride and sodium sulfate

(b) sodium carbonate and calcium bromide

(c) nitrous acid and potassium hydroxide

CHAPTER 13

Chemical Reactions II

How do we write and identify chemical reactions?

What types of reactions are the result of specific types of chemicals?

Learning Outcomes

▪ Identify precipitation, oxidation–reduction, and acid–base reactions.
▪ Write net ionic equations for precipitation and acid–base reactions.
▪ Calculate the oxidation states of various compounds.

Prerequisite Knowledge

▪ Chapter 10: The Periodic Table
▪ Chapter 11: Chemical Nomenclature

Background Information and Questions

This chapter focuses on three main classifications of reactions: precipitation reactions, acid–base reactions, and oxidation–reduction reactions. Let us start with precipitation reactions. They are double or single replacement reactions that occur with ionic compounds. Before we discuss more about precipitation reactions, we need to talk about how ionic compounds behave when dissolved in water. Ionic compounds dissociate into their ions when dissolved in water. Consider the example of table salt (NaCl); table salt is water soluble. When dissolved in water or as salt water, it no longer has the form NaCl. It forms two ions, the sodium ion (Na^+) and the chloride ion (Cl^-).

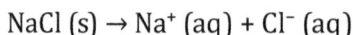

$$NaCl\ (s) \rightarrow Na^+\ (aq) + Cl^-\ (aq)$$

Certain ionic compounds are not soluble in water. For example, silver chloride (AgCl) is not soluble in water. It does not dissociate into its ions. Notice that AgCl remains a solid when dissolved in water; this is called a precipitate.

$$AgCl \text{ (s)} \rightarrow AgCl \text{ (s)}$$

Precipitation reactions are single- or double-replacement reactions and always form a product that is insoluble in water. The following is an example of a balanced precipitation reaction. Two water-soluble compounds ($AgNO_3$ and NaCl) are combined, and they form an insoluble product (AgCl). This is called the *balanced molecular equation.*

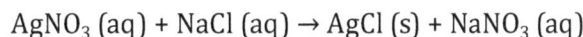

$$AgNO_3 \text{ (aq)} + NaCl \text{ (aq)} \rightarrow AgCl \text{ (s)} + NaNO_3 \text{ (aq)}$$

The balanced molecular equation includes quite a few compounds; sometimes, we like to simplify this equation to illustrate what chemical is being transformed during the reaction. Basically, we only want to write what is reacting; this is called writing the *net ionic equation.* Let's begin with the first step; remember, the compounds with (aq) are not actually compounds when dissolved in water. They dissociate into ions and look like the following equation. Compounds with the (s) designation are not dissolved and do not dissociate into ions. This is known as the *balanced ionic equation.*

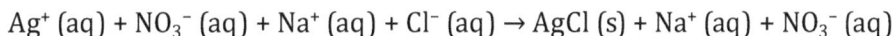

$$Ag^+ \text{ (aq)} + NO_3^- \text{ (aq)} + Na^+ \text{ (aq)} + Cl^- \text{ (aq)} \rightarrow AgCl \text{ (s)} + Na^+ \text{ (aq)} + NO_3^- \text{ (aq)}$$

For the next step, we need to eliminate ions that appear on both sides of the reaction arrow. These are called *spectator ions.* In the balanced ionic equation, both Na^+ and NO_3^- appear on both sides of the arrow. We will cross them out in the following equation:

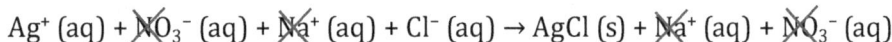

$$Ag^+ \text{ (aq)} + \cancel{NO_3^-} \text{ (aq)} + \cancel{Na^+} \text{ (aq)} + Cl^- \text{ (aq)} \rightarrow AgCl \text{ (s)} + \cancel{Na^+} \text{ (aq)} + \cancel{NO_3^-} \text{ (aq)}$$

The remaining ions and compound(s) constitute what is known as the *net ionic equation.* This is what is "actually happening" during the reaction.

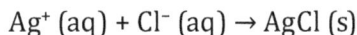

$$Ag^+ \text{ (aq)} + Cl^- \text{ (aq)} \rightarrow AgCl \text{ (s)}$$

Net ionic equations are typically taught alongside precipitation reactions, which result in solid (s) products, but they can also apply to reactions with liquid (l) or gaseous (g) products.

1. What is a spectator ion?

2. Why do we cross out or not include spectator ions in a net ionic equation?

The next type of reaction we will talk about is an *acid–base*, or *neutralization, reaction*. These are just like the name sounds, an acid plus a base as reactants. A good mnemonic to help you remember is Acid + Base → Salt + Water. The following is an example of an acid–base reaction:

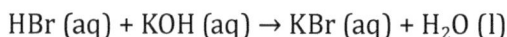

$$HBr \ (aq) + KOH \ (aq) \rightarrow KBr \ (aq) + H_2O \ (l)$$

You can see why this is called a *neutralization reaction*. The acid is neutralized by the base and vice versa.

3. Write a net ionic equation for a neutralization reaction for the HBr and KOH reaction above.

The final type of reaction we will discuss is the *oxidation–reduction*, or *redox, reaction*. These are identified by changes in oxidation state or oxidation number as a reaction occurs. What is an oxidation state, you say? For ions, the oxidation number is simply the charge on the ion. For covalent compounds, the oxidation state is a made-up charge that is assigned to elements to determine how many electrons they have gained or lost. To identify a redox reaction, you have to be able to calculate oxidation numbers. Let us do that first. **Table 13.1** provides some rules to follow when calculating oxidation numbers.

TABLE 13.1 Rules When Calculating Oxidation Numbers

	Rule	Example
1	The oxidation state of an element in its natural state is always zero.	H_2, Cl_2, Mg (s), Zn (s), Fe (s)
2	The oxidation state of a monoatomic ion is always the same as its charge.	Na^+, Mg^{2+}, Fe^{3+}, N^{3-}, Cl^-

(continued)

TABLE 13.1 Rules When Calculating Oxidation Numbers (*Continued*)

	Rule	Example
3	The charge of a polyatomic ion is the sum of all its elements' oxidation numbers.	$CO_3{}^{2-}$, $NO_3{}^-$, $SO_4{}^{2-}$, $S_2O_3{}^{2-}$
4	The sum of all the elements' oxidation numbers in a neutral compound is zero.	$NaCl$, CO_2, SF_6, N_2O, $CaCO_3$
5	Oxygen usually has an oxidation number of –2, except when it is in a peroxide, in a superoxide, or in a compound with only fluorine such as OF_2. In a peroxide, oxygen is a –1. In a superoxide, it is the dianion, $O_2{}^-$.	H_2O vs. H_2O_2
6	Hydrogen usually has an oxidation number of +1, except when it is in a hydride compound. Then it is a –1.	HCl vs. CaH_2

Let us calculate the oxidation number of each element in $CuSO_4$. Copper(II) sulfate is an ionic compound that is composed of two ions: Cu^{2+} and $SO_4{}^{2-}$. Using the rules in Table 13.1, we can identify the oxidation state of Cu; it is 2+. We can also determine that oxygen is a 2⁻ from the rules. One might assume that the sulfur is a 2⁻, since it is in Group 6A on the periodic table; however, that is incorrect because the sulfur is bound to four oxygens. In order to calculate the oxidation number of sulfur, we need to use algebra:

$$SO_4^{2-}$$

$$1\left(x\right)+4\left(-2\right)=\ -2$$
$$x+\ -8=\ -2$$
$$x=\ +6$$

Here we are using x to solve for the oxidation number of one sulfur atom. Because there is one sulfur atom and an unknown oxidation state, the term representing sulfur in the sulfate formula is $1x$. Oxygen's oxidation number is always 2⁻, and there are four oxygen atoms. The term representing oxygen is $4(-2)$, meaning there are four oxygens that each have a –2 oxidation state. Sulfate has an overall charge of –2. Remember that all oxidation states in a polyatomic sum to its charge; hence, we set the entire equation equal to –2. If we solve for x, we find that the oxidation state of the sulfur atom is a 6+.

4. Write the algebraic equation to solve for the oxidation number of carbon in oxalate ($C_2O_4{}^{2-}$).

Now let us look at redox reactions. In redox reactions, two elements' oxidation numbers change. One element undergoes oxidation, while another undergoes reduction. Oxidation is the loss of electrons by an element, whereas reduction is the gain of electrons. One way to remember this is "LEO goes GER." LEO is an acronym for "Loses Electrons Oxidation," and GER stands for "Gains Electrons Reduction." Oxidation and reduction occur in sync; one cannot occur without the other. If an atom is losing electrons, another atom must be there to accept or gain them. In order to identify an oxidation–reduction reaction, you have to calculate the oxidation numbers of every element to determine if there is a change. Let's look at the following equation:

$$2 \; Cu \; (s) + O_2 \; (g) \rightarrow Cu_2O \; (s)$$

In the equation, both the Cu (s) and O_2 (g) are in their natural state, meaning their oxidation numbers are zero. After the reaction, the oxidation numbers have changed. Cu changes from a zero to a 1+, while oxygen changes from a zero to a 2^-. Copper becomes more positive, so it is oxidized and has lost one electron. I remember the following phrase to help me identify oxidation and reduction: "If you lose something negative, you become more positive." Oxygen gains two electrons, or is reduced, and its oxidation number becomes more negative. I remember this by the following statement: "If you gain something negative, you become more negative."

5. Why must oxidation and reduction occur simultaneously?

See **Table 13.2** for a summary of the three main types of reactions discussed in this chapter.

TABLE 13.2 Summary of the Reactions

Reaction Type	Identifier(s)
Precipitation	Reactants = two soluble (aq) ionic compounds. Product = an insoluble (s) ionic compound.
Acid–base	Reactants = one acid and one base. Products = a salt (ionic compound) and water.
Oxidation–reduction	Reactants = ionic, molecular compounds, and/or elements. Changes in oxidation state or number.

Review Questions

6. What are the defining factors that allow you to identify precipitation, acid–base, and oxidation–reduction reactions?

7. Why do we write net ionic equations? Why do you think they are important?

8. Explain why combustion reactions $(C_xH_y + O_2 \rightarrow CO_2 + H_2O)$ might be classified as oxidation–reduction reactions.

Application Questions

9. Nitrous oxide and sulfur oxides can combine with water vapor in the air to form nitric and sulfuric acids. These are the main components of acid rain. Many countries are using lime (CaO or $Ca(OH)_2$) to neutralize rivers and streams that have become too acidic to support life. Write two equations that predict the neutralization between nitric acid and slaked lime ($Ca(OH)_2$) and sulfuric acid and slaked lime.

10. Calculate the oxidation number for each element in the following compounds.

(a) $KMnO_4$

(b) $S_2O_3^{2-}$

(c) N_2O_5

(d) $(NH_4)_2SO_4$

(e) PbS_2

11. Determine which element is oxidized and which is reduced in the following redox equations.

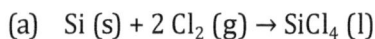

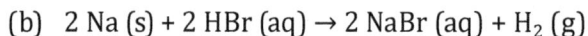

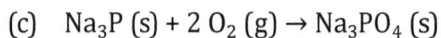

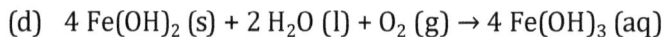

(a) $Si\ (s) + 2\ Cl_2\ (g) \rightarrow SiCl_4\ (l)$

(b) $2\ Na\ (s) + 2\ HBr\ (aq) \rightarrow 2\ NaBr\ (aq) + H_2\ (g)$

(c) $Na_3P\ (s) + 2\ O_2\ (g) \rightarrow Na_3PO_4\ (s)$

(d) $4\ Fe(OH)_2\ (s) + 2\ H_2O\ (l) + O_2\ (g) \rightarrow 4\ Fe(OH)_3\ (aq)$

12. Determine whether the following equations are redox, acid–base, or precipitation reactions.

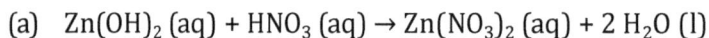

(a) $Zn(OH)_2\ (aq) + HNO_3\ (aq) \rightarrow Zn(NO_3)_2\ (aq) + 2\ H_2O\ (l)$

(b)   Zn (s) + 2 HCl (aq) → ZnCl$_2$ (aq) + H$_2$ (g)

(c)   CaCO$_3$ (aq) + 2 HBr (aq) → CaBr$_2$ (aq) + CO$_2$ (g) + H$_2$O (l)

(d)   MgCl$_2$ (aq) + 2 NaOH (aq) → Mg(OH)$_2$ (s) + 2 NaCl (aq)

(e)   10 I$^-$ (aq) + 2 MnO$_4^-$ (aq) + 8 H$_2$O (l) → 5 I$_2$ (s) + 2 Mn^{2+} (aq) + 16 OH$^-$ (aq)

CHAPTER 14

Introduction to Stoichiometry

How do the number of reactants relate to the number of products?

What is a mole in chemical terms?

Learning Outcomes

- Demonstrate understanding of the term "mole."
- Calculate the amount of substance produced by a chemical reaction.

Prerequisite Knowledge

- Chapter 1: Math for Chemists
- Chapter 3: Atomic Structure
- Chapter 12: Chemical Reactions I
- Chapter 13: Chemical Reactions II

Background Information and Questions

Stoichiometry involves the calculation of amounts of products and reactants in a chemical reaction. The first step is to have a balanced chemical equation (covered in the previous lesson), so let's start with that.

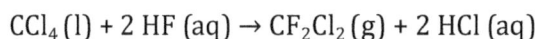

$$CCl_4 \text{ (l)} + 2\ HF \text{ (aq)} \rightarrow CF_2Cl_2 \text{ (g)} + 2\ HCl \text{ (aq)}$$

Previously, you learned that the coefficients in a balanced chemical equation represent the number of molecules (or atoms) that are participating. This is true, but in reality there will be many molecules all undergoing the reaction at the same time. We represent the total number of molecules using the mole (mol). One mole of a substance is equal to 6.022×10^{23} molecules (or atoms) of that substance. This number is known as Avogadro's number, and a mole of anything will contain this number of somethings (typically molecules or atoms), just as a dozen always contains 12 of something. For example, in the above equation we see 1 mole of CCl_4 reacting with 2 moles of HF to produce 1 mole of CF_2Cl_2 and 2 moles of HCl. The 1 mole of CCl_4 contains 1 mole of carbon and 4 moles of chlorine.

The following is a balanced equation for the formation of rust:

$$2 \ Fe \ (s) + O_2 \ (g) \rightarrow 2 \ FeO \ (s)$$

1. How many moles of iron(II) oxide are formed from 2 moles of Fe and 1 mole of O_2?

2. How many moles of iron(II) oxide would be formed from 4 moles of Fe and 2 moles of O_2?

3. How many molecules of iron(II) oxide are present in the number of moles from question 2?
    Recall that there are 6.022×10^{23} atoms/molecules in a mole.

4. In 2.89×10^{24} molecules of O_2, how many moles are present?

The coefficients in chemical reactions also serve as a ratio between reactants and products. This allows you to determine the amount of products a reaction will produce based on the amount of reactants you begin with. For example:

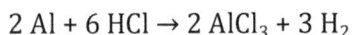

$$2 \ Al + 6 \ HCl \rightarrow 2 \ AlCl_3 + 3 \ H_2$$

If we begin this reaction with 7 moles of aluminum and an unlimited amount of hydrochloric acid, how much hydrogen gas will be produced in moles? (*Hint: Reviewing dimensional analysis from Chapter 1 might be useful here!*)

$$\frac{7 \text{ mol Al}}{1} \times \frac{3 \text{ mol H}_2}{2 \text{ mol Al}} = \text{ ?}$$

$$\frac{7 \text{ mol Al}}{1} \times \frac{3 \text{ mol H}_2}{2 \text{ mol Al}} = 10.5 \text{ mol H}_2$$

This calculation begins with the number of moles that are participating in the reaction (7). The coefficients from the chemical equation appear in the second part of this calculation as a molar ratio (3/2)—these ratios between different molecules must always be done in moles. Units "cancel out" just as in previous chapters and leave us with the amount of hydrogen that would be produced as a product.

5. Using the aluminum and hydrochloric acid reaction, how many moles of $AlCl_3$ would be produced in a reaction that began with 12 moles of HCl?

6. How many moles of HCl would be required to react with 4 moles of Al?

Moles are extremely useful for chemists, but we cannot measure them directly in the way that we can measure grams. Fortunately, it is simple to convert from moles to grams (and vice versa) using the molecular weight of the compound. The molecular weight is the weight in grams of 1 mole of a substance; it has the same numerical value as the atomic mass (see Chapter 3).

For example, the atomic mass of $AlCl_3$ is 133.34 amu; therefore, the molar mass is 133.34 g/mol. In other words, 1 mole of $AlCl_3$ will weigh 133.34 grams. We can use this as a conversion factor to convert between grams and moles.

How many moles are present in 400 grams of $AlCl_3$? The answer is below!

$$\frac{400 \text{ g AlCl}_3}{1} \times \frac{1 \text{ mol AlCl}_3}{133.34 \text{ g AlCl}_3} = 3 \text{ mol AlCl}_3$$

7. How many moles of HCl are present in 182 g of HCl?

8. How many grams of HCl are present in 20 moles of HCl?

This conversion can be used to directly calculate the number of grams (or moles) that are present as reactants or products in a reaction. For example:

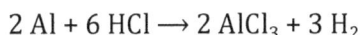

$$2 \text{ Al} + 6 \text{ HCl} \longrightarrow 2 \text{ AlCl}_3 + 3 \text{ H}_2$$

How many grams of $AlCl_3$ would be produced from 5 moles of HCl and unlimited Al? (*Hint: The coefficients in the chemical equation must be used for the ratio between different compounds!*)

$$\frac{5 \text{ mol HCl}}{1} \times \frac{2 \text{ mol AlCl}_3}{6 \text{ mol HCl}} \times \frac{133.34 \text{ g AlCl}_3}{1 \text{ mol AlCl}_3} = \; ?$$

$$\frac{5 \text{ mol HCl}}{1} \times \frac{2 \text{ mol AlCl}_3}{6 \text{ mol HCl}} \times \frac{133.34 \text{ g AlCl}_3}{1 \text{ mol AlCl}_3} = 222 \text{ g AlCl}_3$$

9. How many grams of H_2 would be produced from 2 moles of HCl and an unlimited amount of Al in the above equation?

The most common form of stoichiometry is gram-to-gram stoichiometry, wherein both the starting amount of reactant and the final amount of product are listed in grams instead of moles. *Remember that you must always convert to moles when doing stoichiometry! It is moles, not grams, that allow for the ratio of products to reactants.*

How many grams of $AlCl_3$ could be produced from 50 g of HCl and an unlimited amount of Al? To find this, you use the same balanced equation from the previous example:

$$\frac{50 \text{ g HCl}}{1} \times \frac{1 \text{ mol HCl}}{36.5 \text{ g HCl}} \times \frac{2 \text{ mol AlCl}_3}{6 \text{ mol HCl}} \times \frac{133.34 \text{ g AlCl}_3}{1 \text{ mol AlCl}_3} = \; ?$$

$$\frac{50 \text{ g HCl}}{1} \times \frac{1 \text{ mol HCl}}{36.5 \text{ g HCl}} \times \frac{2 \text{ mol AlCl}_3}{6 \text{ mol HCl}} \times \frac{133.34 \text{ g AlCl}_3}{1 \text{ mol AlCl}_3} = 60.9 \text{ g AlCl}_3$$

10. How many grams of H_2 could be produced from 50 g of HCl and an unlimited amount of Al?

11. How many grams of HCl would be required to react completely with 100 g of Al?

Review Questions

12. A chemical reaction involves 2 moles of A reacting with 3 moles of B to produce 4 moles of C. If this reaction starts with 6 moles of A and 9 moles of B, how many moles of C will be produced?

13. How many doughnuts are in a mole of doughnuts?

14. You have a sample of copper and a sample of zinc. Each sample contains the same number of atoms. Which sample weighs more?

Application Questions

15. How many carbon atoms are in 50.0 g of carbon tetrachloride?

16. How many chlorine atoms are in 50.0 g of carbon tetrachloride?

17. How many hydrogen atoms are in 50.0 g of carbon tetrachloride?

18. How many moles of copper will be required to react with 2.5 moles of nitric acid, given the following reaction?

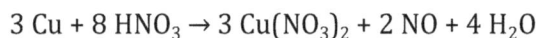

$$3 \text{ Cu} + 8 \text{ HNO}_3 \rightarrow 3 \text{ Cu(NO}_3)_2 + 2 \text{ NO} + 4 \text{ H}_2\text{O}$$

19. Using the reaction from the previous question, how many moles of water will be formed if the reaction begins with 50.0 g Cu and unlimited HNO_3?

20. How many grams of NO will be formed if the reaction in the previous question starts with 25.6 g of nitric acid and unlimited Cu?

21. If 70.0 g of F_2 and Xe react, how many grams of XeF_6 can be formed?

$$Xe + 3\ F_2 \rightarrow XeF_6$$

CHAPTER 15

Limiting Reactants and Molarity

Is it possible to run out of one reactant in a reaction?

How do chemists determine the concentration of a solution?

Learning Outcomes

- Identify the limiting reagent in a chemical reaction.
- Calculate molarity, number of moles, or volume of solution if given two of these three values.
- Calculate the number of grams required for a solution of a given molarity.
- Perform dilution calculations.

Prerequisite Knowledge

- Chapter 14: Introduction to Stoichiometry

Background Information and Questions

In the previous chapter, we practiced stoichiometry using the assumption that one of our reactants was in excess, meaning that we could not run out of that reactant. But what if we only have a limited amount of both reactants? In that case one of the reactants will be termed the *limiting reactant* because it will limit the amount of product that can be produced. In other words, the limiting reactant will run out first.

Consider the reaction from the previous chapter:

$$2 \text{ Al} + 6 \text{ HCl} \rightarrow 2 \text{ AlCl}_3 + 3 \text{ H}_2$$

If this reaction began with 2 moles of Al and 7 moles of HCl, the Al would run out first and leave 1 mole of HCl in excess because there is no more Al for it to react with—the Al is the limiting reactant. Similarly, if this reaction began with 4 moles of Al and 9 moles of HCl, the HCl would run out first and leave 1 mole of Al behind—the HCl is the limiting reactant. The limiting reactant therefore is not the reactant of which you have the smallest amount, nor the one with the smallest coefficient, but rather the reactant that is used up first.

Consider the following reaction:

$$1\,N_2 + 3\,H_2 \rightarrow 2\,NH_3$$

1. If this reaction begins with 2.0 moles of nitrogen and 5.0 moles of hydrogen, which is the limiting reactant? Which is in excess?

2. If this reaction begins with 2.0 moles of nitrogen and 8.0 moles of hydrogen, which is the limiting reactant? Which is in excess?

One of the most straightforward ways to solve a more advanced limiting reagent problem is to simply work the problem twice, once for each reactant. The reactant that produces the least amount of product will be the limiting reactant. This amount of product is the actual amount that will be produced. Let's work an example. How much NH_3 will be produced if 500 g of N_2 react with 400 g of H_2 in the following reaction?

$$1\,N_2 + 3\,H_2 \rightarrow 2\,NH_3$$

This is a gram-to-gram stoichiometry problem (see the previous chapter), but in this case we do not know which reactant is limiting and which is in excess. Therefore, we work the problem twice, once for each reactant.

Assuming that the hydrogen is unlimited and the 500 g of nitrogen limit the reaction:

$$\frac{500\text{ g }N_2}{1} \times \frac{1\text{ mol }N_2}{28\text{ g }N_2} \times \frac{2\text{ mol }NH_3}{1\text{ mol }N_2} \times \frac{17\text{ g }NH_3}{1\text{ mol }NH_3} = 607.1\text{ g }NH_3$$

Assuming that the nitrogen is unlimited and the 400 g of hydrogen limit the reaction:

$$\frac{400 \text{ g H}_2}{1} \times \frac{1 \text{ mol H}_2}{2 \text{ g H}_2} \times \frac{2 \text{ mol NH}_3}{3 \text{ mol H}_2} \times \frac{17 \text{ g NH}_3}{1 \text{ mol NH}_3} = 2{,}266.7 \text{ g NH}_3$$

Compare the results from these two calculations. When we assume that the N_2 is our limiting reagent for the calculation, we get 607.1 g NH_3 as our product. When we assume that the H_2 is our limiting reagent for the calculation, we get 2,266.7 g NH_3 as our product. Therefore, our limiting reagent is the N_2; that is, when mixing 500 g of nitrogen and 400 g of hydrogen, we would run out of nitrogen, have excess hydrogen, and produce approximately 607 g of ammonia. This is called the theoretical yield.

Consider the following reaction:

$$2 \text{ Al} + 6 \text{ HCl} \rightarrow 2 \text{ AlCl}_3 + 3 \text{ H}_2$$

3. If 150.0 g of Al react with 200.0 g of HCl, how many grams of $AlCl_3$ will be produced?

4. Which is the limiting reactant, Al or HCl?

Most—but not all—chemical reactions are done in solution. In other words, reactions are often performed by dissolving solid reagents (solutes) in water (the solvent) to form a solution (*solutions in general will be covered in more detail in Chapter 20*). In these cases it is vital to understand the concept of concentration.

The most common unit of concentration in chemistry is molarity (M). Molarity is the number of moles of solute in exactly 1 L of solution.

$$\text{Molarity} = \frac{(\text{mol solute})}{(\text{L solution})}$$

$$\text{M} = \frac{\text{mol}}{\text{L}}$$

5. A 1.00 L sample of salt water contains 3.00 moles of NaCl. What is the molarity?

6. A 500.0 mL sample of salt water contains 2.00 moles of NaCl. What is the molarity?

7. A 500.0 mL sample of salt water contains 20.0 g of NaCl. What is the molarity? *Hint: Recall how to convert grams into moles!*

Solutions can be diluted to form solutions that are less concentrated. The standard equation for dilution is:

$$M_1V_1 = M_2V_2$$

in which M_1 and V_1 represent the molarity and volume of the original/concentrated solution and M_2 and V_2 represent the molarity and volume of the resulting dilute solution. Consider the following examples:

10 mL of a 4.00 M solution of NaCl is diluted to 50 mL. What is the new concentration?

$$M_1V_1 = M_2V_2$$

$$(4.00 \text{ M})(10 \text{ mL}) = (M_2)(50 \text{ mL})$$

$$M_2 = 0.8 \text{ M NaCl}$$

How many mL of a 6.0 M solution should be diluted to obtain 100 mL of a 1.5 M solution?

$$M_1V_1 = M_2V_2$$

$$(6.0 \text{ M})(V_1) = (1.5 \text{ M})(100 \text{ mL})$$

$$V_1 = 25 \text{ mL}$$

8. 40.0 mL of a 6.00 M solution of NaCl is diluted to 200.0 mL. What is the new concentration?

9. How many milliliters of a 2.0 M solution should be diluted to obtain 50.0 mL of a 0.50 M solution?

10. 20.0 mL of a 3.5 M solution is diluted to a concentration of 1.0 M. What is the final volume?

Review Questions

11. A chemical reaction involves 2 moles of A reacting with 3 moles of B to produce 4 moles of C. If this reaction starts with 6 moles of A and an unlimited amount of B, how many moles of C will be produced?

12. A chemical reaction involves 2 moles of A reacting with 3 moles of B to produce 4 moles of C. If this reaction starts with 6 moles of A and 6 moles of B, how many moles of C will be produced?

13. Is dilution a reversible process as described in this chapter? Explain your reasoning for whatever answer you choose!

Application Questions

14. Answer the following questions.

    (a) If the following reaction is carried out using 20.0 g of xenon and 20.0 g of fluorine, which is the limiting reactant?

$$Xe + 3\ F_2 \rightarrow XeF_6$$

    (b) How many grams of xenon hexafluoride would be produced from the reaction?

    (c) Which reactant is left over—and how much of it is present—after the reaction is complete?

15. A beaker containing 50.0 mL of 0.40 M HCl is diluted to a concentration of 0.15 M. How much water was added?

16. You need to prepare 200.0 mL of a solution that is 2.00 M H_2SO_4 from a solution that is 12.0 M H_2SO_4. How much of the concentrated solution will you need to dilute?

CHAPTER 16

Stoichiometry Continued

Is there more than one type of chemical formula?

What happens if you get less product than the stoichiometry says you should?

Learning Outcomes

- Calculate the theoretical, actual, and percent yields of a reaction.
- Convert between molecular and empirical formulas.
- Demonstrate an ability to perform a combustion analysis.

Prerequisite Knowledge

- Chapter 15: Limiting Reactants and Molarity

Background Information and Questions

The yield, or amount of product, that we have calculated in the previous chapters is known as the *theoretical yield*. In theory, this is the amount of product that should, mathematically, be produced in a chemical reaction. However, a real experiment in a laboratory setting will often produce less than the amount calculated in theory. The amount of product actually produced is known as the *actual yield*. The *percent yield*, which we will focus on in this chapter, is the actual yield divided by the theoretical yield times 100.

$$\text{Percent yield} = \frac{\text{actual yield}}{\text{theoretical yield}} \times 100$$

In the previous chapter, we calculated that 500 g of nitrogen and 400 g of hydrogen *should* produce approximately 607 g of ammonia. Therefore, the 607 g NH_3 is our theoretical yield. However, when we physically perform this experiment in the lab, we only obtain 550 g NH_3, which is our actual yield. What is the percent yield in this example?

$$\text{Percent yield} = \frac{550 \text{ g}}{607 \text{ g}} \times 100 = 82\% \text{ yield}$$

1. If the theoretical yield for a given reaction is 400.0 g and the actual yield is 388.0 g, what is the percent yield?

2. Consider the following reaction:

$$2 \text{ Al} + 6 \text{ HCl} \rightarrow 2 \text{ AlCl}_3 + 3 \text{ H}_2$$

If 50.0 g of HCl react with an excess of Al to produce 55.0 g of $AlCl_3$, what is the percent yield?

Hint: First calculate the theoretical yield—how much $AlCl_3$ "should" you get?

We can also use stoichiometry to determine the formulas of chemical compounds. Thus far we have used molecular formulas when dealing with chemical formulas. A molecular formula provides the exact number of atoms of each element in a molecule using subscripts. The empirical formula is another type of chemical formula that provides the smallest ratio of elements instead of the exact number. To convert from the molecular formula to the empirical formula, simply divide by the lowest common denominator for the subscripts:

Molecular formula: $C_6H_{12}O_6$ Molecular formula: N_2O_4

Empirical formula: CH_2O Empirical formula: NO_2

3. What is the empirical formula for C_2H_6?

4. What is the empirical formula for C_8H_{18}?

Sometimes a chemical experiment can only provide information on the percentage of each element in a compound. In this case it is still possible to determine the empirical formula by looking at the molar ratio. For example, if a compound contains 30.4% nitrogen and 69.6% oxygen by mass, what is the empirical formula?

Start by assuming that you have a 100 g sample. This allows us to convert the percent to grams, which can then be converted into moles.

$$30.4\% \text{ nitrogen in a 100 g sample} = 30.4 \text{ g nitrogen} \rightarrow 2.17 \text{ mol N}$$

$$69.6\% \text{ oxygen in a 100 g sample} = 69.6 \text{ g oxygen} \rightarrow 4.35 \text{ mol O}$$

Find the lowest whole number ratio for these numbers. These values will be the subscripts in the empirical formula. The ratio of 4.35 to 2.17 is 2:1; therefore, there are two oxygens for every nitrogen in the empirical formula, NO_2.

5. A compound is determined to be 27% sulfur, 13.4% oxygen, and 59.6% chlorine. What is the empirical formula?

Combustion analysis is a method by which the composition of hydrocarbons (compounds consisting primarily of hydrogen and carbon) can be determined. In combustion analysis a sample is heated with oxygen gas until all of the C and H has reacted to form CO_2 and H_2O. These gases can be weighed to determine the original amount of C and H in the sample, which can then be used to determine the empirical formula.

For example, say that we have a sample of an unknown hydrocarbon that undergoes combustion analysis, resulting in 5.40 g H_2O and 8.80 g CO_2. What is the empirical formula? Our goal in determining the empirical formula is to determine the molar ratio between carbon and hydrogen. To do this we must first determine the number of moles of H and C present then find the lowest whole number ratio.

$$H: \frac{5.40 \text{ g } H_2O}{1} \times \frac{1 \text{ mol } H_2O}{18 \text{ g } H_2O} \times \frac{2 \text{ mol H}}{1 \text{ mol } H_2O} = 0.6 \text{ g H}$$

$$C: \frac{8.80 \text{ g } CO_2}{1} \times \frac{1 \text{ mol } CO_2}{44 \text{ g } CO_2} \times \frac{1 \text{ mol C}}{1 \text{ mol } CO_2} = 0.2 \text{ g C}$$

The ratio of C to H is 1:3; therefore, our empirical formula is CH_3.

6. What is the empirical formula of a compound that undergoes combustion analysis to produce 0.34 g H_2O and 1.7 g CO_2?

Review Questions

7. Why should you be suspicious if your actual yield is larger than your theoretical yield?

8. Going from a molecular formula to an empirical formula is easy. What additional information would you require to go from an empirical formula to a molecular formula?

9. Use the equation A + B → C to answer the following questions. Stoichiometry predicts that 20 g of A and 30 g of B should produce 40 g of C. However, when you perform the experiment, you obtain 30 g of C.

   (a) What is the theoretical yield?

   (b) What is the actual yield?

(c) What is the percent yield?

Application Questions

10. Oxygen (O_2) and propane (C_3H_8) react to form carbon dioxide (CO_2) and water (H_2O). If 40.0 g of carbon dioxide are produced from 20.0 g of propane and unlimited oxygen, what is the percent yield?

Hint: Yes, you do need to write out and balance the chemical reaction first!

11. A compound contains 29.0% sodium, 30.0% oxygen, and 41.0% sulfur. What is the empirical formula of this compound?

12. A 0.500 g sample of hydrocarbon (only carbon and hydrogen) undergoes combustion to produce 1.69 g of carbon dioxide and 0.346 g of water. What is the empirical formula of this compound?

13. A compound that contains C, H, *and* O undergoes combustion. After burning a 1.038 g sample of this compound, you obtain 2.48 g carbon dioxide and 0.510 g water. What is the empirical formula of this compound?

CHAPTER 17

Gases

What are different ways to report pressure?

What parameters of a gas can we calculate?

What is the theory behind gas behavior?

Learning Outcomes

- Perform pressure calculations.
- Rationalize each gas law and recognize when to utilize each to calculate gas parameters.
- Describe kinetic molecular theory (KMT).

Prerequisite Knowledge

- Chapter 2: Matter
- Chapter 14: Introduction to Stoichiometry

Background Information and Questions

Gases are a significant phase of matter; they are essential for life and historically played an important role in the development of chemistry. Oxygen and carbon dioxide are both necessary for biological processes on Earth, while nitrogen is the gas responsible for airbag inflation. The study of gas behavior is essential for chemists. Within a certain range of parameters, gases behave in a certain predictable manner. This predictable manner can be modeled on four parameters: pressure, temperature, volume, and amount.

Pressure is a measure of a force exerted over an area. In a gas, pressure (P) is exerted by the gas on the walls of the container. Pressure is measured in a variety of units.

Here are conversion factors containing many common pressure units that you might encounter in this chapter:

$$760 \ mmHg = 1 \ atm = 14.7 \ psi = 760 \ torr = 101{,}325 \ Pa = 29.9 \ inHg$$

1. Convert the following pressures from one unit to another.

   (a) 747.0 mmHg to atm

   (b) 200.0 torr to mmHg

   (c) 35.0 psi to atm

   (d) 2.00 atm to Pa

   (e) 5000.0 Pa to kPa
   (This one isn't listed above, but what does a "k" typically mean at the front of a unit?)

Volume (V) is the amount of space occupied by the gas. It is typically measured in liters (L) or milliliters (mL). Temperature (T) is a measure of the molecular motion (measured in Kelvin [K]), while amount is the number of gas atoms/molecules (n, measured in moles). These parameters are utilized to model gas behavior through a series of gas laws:

I. *Boyle's law*: Robert Boyle theorized this law, which states that pressure and volume are inversely proportional if temperature and amount are held constant. The law is symbolized as follows:

$$P_1 V_1 = P_2 V_2$$

II. *Charles's law*: First hypothesized by Jacques Charles, the law states that volume and temperature are directly proportional if amount and pressure are held constant. This is represented by the following equation:

$$\frac{V_1}{T_1} = \frac{V_2}{T_2}$$

III. *Avogadro's law (Avogadro's hypothesis)*: Amedeo Avogadro determined that the number of molecules of a gas is directly proportional to the volume if pressure and temperature are held constant:

$$\frac{V_1}{n_1} = \frac{V_2}{n_2}$$

IV. *Amontons's law (Gay-Lussac's law)*: French physicist Guillaume Amontons discovered that if volume and the number of molecules are held constant, then pressure and temperature are directly proportional. This law is also credited to Joseph Louis Gay-Lussac, who also discovered the law of combining volumes. This relationship is represented by the following equation:

$$\frac{P_1}{T_1} = \frac{P_2}{T_2}$$

V. *Combined gas law*: The combined gas law combines Boyle's, Charles's, and Amontons's laws into one equation that represents the relationships of pressure, volume, and temperature for a gas. This equation is useful when more than one parameter is being altered for a gas:

$$\frac{P_1 V_1}{T_1} = \frac{P_2 V_2}{T_2}$$

VI. *Ideal gas law*: The ideal gas law fuses Boyle's, Charles's, and Avogadro's laws. Gases only obey the law if certain assumptions, which are discussed later in this chapter, are made. With those assumptions, the following equation can be used to model an ideal gas:

$$PV = nRT$$

The ideal gas law can also be utilized to determine various other aspects about gases, such as density (d) and molar mass (M). The following equation for the density of a gas is derived from the ideal gas law:

$$d = \frac{PM}{RT}$$

2. Before an experiment, your gas is stable at a pressure of 0.80 atm and 300.0 K. After heating this gas to 500.0 K, what is the pressure?

3. Does the previous question represent a direct or an indirect relationship between pressure and volume? Explain.

4. Compare the two equations listed for the ideal gas law (Part VI under gas laws). Assuming that density is mass over volume, how would you derive the second equation from the first?

5. If you knew the density of a gas, which two additional variables would you need to determine the molar mass of the gas?

The assumptions made for the ideal gas law come from the *kinetic molecular theory* (KMT) for gases. This model is used to predict the behavior of gases as well as explain the various gas laws described previously. The following is the list of assumptions of the KMT:

I. Gaseous atoms or molecules have no volume but do have a mass. In other words, the gas particles' volume is tiny compared to the volume of their container and the distances between molecules. The volume of each molecule is therefore ignored.

II. Gas particles do not experience attractive or repulsive forces between themselves. This assumption simplifies the discussion of energy by disregarding the potential energy associated with intermolecular attraction or repulsion. In other words, the assumption indicates that gas particles only possess kinetic energy.

III. Gaseous atoms or molecules have continuous, random motion.

IV. Collisions between particles and the sides of the container are elastic. There is no loss of energy when a collision occurs; kinetic energy is transferred.

V. The average kinetic energy of a gas, regardless of its identity, is proportional to the temperature in Kelvin. This means molecular motion ceases at absolute zero.

6. Refer to the following figure. Redraw the illustration indicating a twofold pressure increase.
   (Note: There is more than one way to do this!)

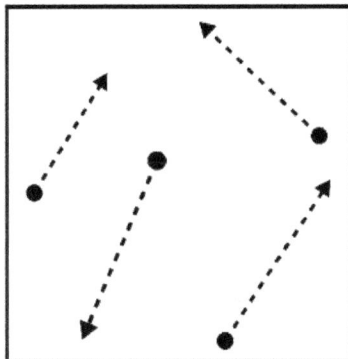

7. Draw two illustrations that best illustrate Avogadro's law.

Let's apply the KMT to the first two gas laws:

   Boyle's law: Decreasing the volume of a container increases the pressure. This is because decreasing the volume decreases the surface area of the container. A lower surface area increases the pressure, $P = \frac{Force}{Area}$. You can also think about this from the perspective that the same number of molecules collide with the walls of the container more because the container is smaller.

   Charles's law: Increasing the temperature at constant pressure results in an increase in volume. This can be explained by the KMT by stating that an increase in temperature results in an increase in molecular motion; therefore, the volume must increase to maintain the constant pressure.

8. Utilize the KMT to describe Amontons's law at the molecular level.

Review Questions

9. Why is temperature always measured in Kelvin in reference to a gas?

10. Why do we need assumptions when describing the relationships within the ideal gas law?

11. Draw a graph illustrating Charles's law.

Application Questions

12. Convert the following pressures to atmospheres.

    (a) 734 mmHg

    (b) 55.3 psi

    (c) 108.7 kPa

13. 6.39 moles of N_2 occupies a 3.40 L container at 8.10 atm. Calculate the volume of a container needed to hold this amount of N_2 at 5.80 atm.

14. Determine the amount of pressure needed to force 4.76 L of H_2 at 755 mmHg and 25 °C into a smaller 184.0 mL container at 15 °C.

15. Standard temperature and pressure (STP) is sometimes used to denote conditions (273 K and 1 atm) of a gas. Calculate the volume of 1 mole of a gas at STP.

    Note: This is a useful number to memorize!

16. Average male lungs have a capacity of 4.20 L. If the average molar mass of air is 29.0 g/mol and the density is 1.20 g/L, what is the pressure in an average adult female's lungs of a 2.70 L breath of air at body temperature (37 °C)?

17. What is the molar mass of 4.27 g of a gas that occupies 14.3 L at STP?

Figure Credit

Img. 17.1: Source: https://commons.wikimedia.org/wiki/File:Gas_particles_in_a_square.svg.

CHAPTER 18

Advanced Gas Concepts

How do gases work in the real world rather than an ideal one?

How quickly does a delicious smell move through a room?

What happens if gases are mixed at different pressures?

Learning Outcomes

- Identify the conditions in which gases behave in an ideal versus nonideal manner.
- Utilize the van der Waals equation to calculate nonideal gas parameters.
- Relate RMS speed and its factors to rates of diffusion/effusion.
- Explain Dalton's law and utilize it to calculate parameters for mixtures of gases.

Prerequisite Knowledge

- Chapter 17: Gases

Background Information and Questions

Gases behave ideally at high temperatures and low pressures. Real gases deviate from ideal gas behavior at low temperatures and high pressures; these conditions slow particles down and bring them closer together, allowing for more significant attractive and repulsive forces. High pressures typically result in more molecules packed in a space, meaning the particles occupy a significant portion of the volume in the container. These are all conditions that do not align with assumptions of the KMT and the ideal gas law.

1. Conceptually, why do gases behave ideally at high temperatures and low pressures?

In order to correct for the molecular attraction and volume present in real gases, we can alter the ideal gas equation to include two constants: a and b. These are known as the *van der Waals constants*; "a" corrects for intermolecular attractions and repulsions, while "b" corrects for the volume of the gas particles. The resulting equation is known as the *van der Waals equation of state* or the *real gas law*:

$$\left(P + \frac{n^2 a}{V^2}\right)(V - nb) = nRT$$

2. What would be the values of "a" and "b" if a gas behaved ideally?

The molecular motion of gases is related to various properties of the gas. We can calculate the root-mean-square (RMS) speed of a molecule (the square root of the average of the squares of the velocities) by the following equation. Note how temperature and molar mass are the two determining factors. In this equation, the gas constant is 8.3145 J/mol·K; the molar mass will be in units of kg/mol.

$$u_{RMS} = \sqrt{\frac{?RT}{M}}$$

The RMS speed is directly proportional to the rates of diffusion and effusion in gases. Diffusion is the movement of particles from an area of high concentration to an area of low concentration. Effusion is similar to diffusion; however, it is the escape of gases through a small hole into a vacuum. This is illustrated by **Figure 18.1**.

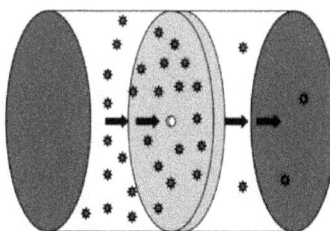

FIGURE 18.1

3. If someone brings a pizza into a room you are in, does the smell travel from the pizza to your nose via effusion or diffusion? Why?

4. Would a higher molar mass cause a gas to effuse faster or slower?

5. Illustrate a gas undergoing diffusion.

To compare the rate of effusion of two different gases, we utilize *Graham's law of effusion*, hypothesized by Thomas Graham, a Scottish chemist. Graham's law states that the rate of diffusion of a gas is equal to the square root of the inverse of its molar mass. The following is an equation relating the rates of effusion of gases A and B:

$$\frac{\text{Rate of effusion A}}{\text{Rate of effusion B}} = \sqrt{\frac{M_B}{M_A}}$$

6. If Gas #1 is four times heavier than another, then according to Graham's law it will effuse through a pinhole at which rate compared to Gas #2? Circle the answer.

Half Double Quadruple Quarter

At this point, we've only discussed modeling the behavior of one type of gas at a time. In the KMT, the identity of the gas particles is irrelevant. A mixture of gases has multiple gas-particle types—the best example of this is air, which contains several different gases, including nitrogen and oxygen. John Dalton hypothesized that in a mixture of gases, the total pressure of the mixture is the sum of the individual pressures of each gas-particle type. This is known as *Dalton's law of partial pressures*.

$$P_{total} = P_1 + P_2 + P_3 + \ldots$$

7. A standard sample of dry air contains roughly 78.0% N, 21.0% O, and 1.0% Ar. At STP, the total pressure of this air sample is 1 atm. What is the pressure, in atm, exerted by the nitrogen (P_{N_2})?

8. The atmosphere on Mars is 95.3% CO_2, 2.7% N_2, 1.6% Ar, 0.13% O_2, and 0.08% CO, and the pressure is only 0.007 atm.[1,2] Which gas would contribute the most and which would contribute the least to the atmospheric pressure on Mars?

Last but not least, let us discuss the stoichiometry of gases. In this chapter and the previous one, we have discussed various ways in which molecular weight, moles, volume, pressure, and temperature relate to each other for gaseous compounds. In chapters 14, 15, and 16 we learned how to obtain the moles of a product generated by a chemical reaction. For gases we can calculate that one mole of any gas at STP will occupy 22.4 L of volume, allowing us to convert from moles to volume with relative ease.

9. Assume 4.00 moles of carbon dioxide gas are produced by the following equation at STP. How many liters of carbon dioxide are produced?

$$CaCO_3 \text{ (s)} \rightarrow CaO \text{ (s)} + CO_2 \text{ (g)}$$

Review Questions

10. When working a gas problem, would the real gas law be a better choice to use than the ideal gas law? Why don't we always use the real gas law?

---

1   Sharp, T. (September, 2017). Mars' Atmopshere: Composition, Climate & Weather. Space.com https://www.space.com/16903-mars-atmosphere-climate-weather.html

2   Schmidt, L. Mars 101: Humans and Atmospheric Pressure. Phoenix Mars Mission. http://phoenix.lpl.arizona.edu/mars103.php

11. Why do all gases occupy 22.4 L at STP? Why doesn't the nature/composition of the gas matter?

Application Questions

12. Calculate the volume of 8.40 moles of Cl_2 (g) at 15.6 atm and -28.0 °C using ideal gas modeling. Do the same calculation with nonideal gas modeling; "a" is 6.260 atm·L^2/mol^2, and "b" is 0.0542 L/mol for Cl_2.

13. What temperature would be required to quadruple the RMS speed of F_2 molecules?

14. Place the following molecules in order of increasing rate of effusion.

   (a) Cl_2, O_2, H_2, Xe

   (b) Rn, F_2, N_2, CH_4

   (c) SO_2, SF_6, Br_2, He

15. An experimental balloon contains gases with a total pressure of 800.0 torr. The gases present are helium (300.0 torr), hydrogen (50.0 torr), and an unknown amount of neon. What is the pressure of the neon gas in the balloon?

16. Assume there are two containers. Container #1 has 20.0 L of oxygen gas at 3.00 atm. Container #2 has 10.0 L of neon gas at 2.00 atm. Both gases are at 300 K. If these gases are mixed and placed into a 5.00 L container, what will be the new total pressure?

17. A mixture of gases containing 24.8 g of F_2 and 32.4 g of SF_6 is cooled to 13.3 °C in a 20.0 L container. Calculate the partial pressure of each gas and the total pressure of the system.

18. When a hydrocarbon such as ethylene (shown below) is burned in oxygen, it forms carbon dioxide and gaseous water. How many liters of water vapor would be formed from 5.00 L of C_2H_4 gas?

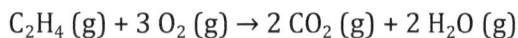

$$C_2H_4 \text{ (g)} + 3\ O_2 \text{ (g)} \rightarrow 2\ CO_2 \text{ (g)} + 2\ H_2O \text{ (g)}$$

19. How many liters of carbon dioxide would be formed if 50.0 g of C_2H_4 gas reacted with an excess of oxygen?

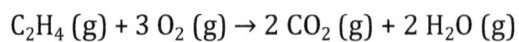

$$C_2H_4 \text{ (g)} + 3 \text{ } O_2 \text{ (g)} \rightarrow 2 \text{ } CO_2 \text{ (g)} + 2 \text{ } H_2O \text{ (g)}$$

Figure Credit

Fig. 18.1: Copyright © by Astrang13 (CC BY-SA 3.0) at https://commons.wikimedia.org/wiki/File:Effusion.svg.

CHAPTER 19

Liquids and Solids

What types of forces occur between molecules?

Why are liquids viscous?

Why are some solids soft and others hard?

Learning Outcomes

- Describe intermolecular forces (IMFs) and determine the type experienced by a molecule.
- Explain the role of IMFs in properties of liquids.
- Rationalize the properties of crystalline versus amorphous solids.
- Utilize unit cells to describe the structure of crystalline solids.

Prerequisite Knowledge

- Chapter 9: Molecular Geometry
- Chapter 10: The Periodic Table

Background Information and Questions

Liquids and solids have different properties than gases. Liquids have a defined volume but an undefined shape. Gases have neither a defined volume nor shape. The particles in liquids and solids are much closer together and have lower kinetic energy than those in gases; this allows the particles to experience attractive and repulsive forces between themselves. These forces are known as *intermolecular forces* (IMFs). Intermolecular forces occur between molecules because of electron-rich (a partial negative charge, δ^-)

and electron-poor regions (a partial positive charge, δ^+). **Table 19.1** describes the different types of IMFs.

TABLE 19.1 Types of Intermolecular Forces

Force	Definition	Example	Relative Strength
Induced dipole–induced dipole (London dispersion)	Involves temporary dipoles between any molecule (typically occurs between nonpolar molecules)	Ex: Ar–Ar	Very weak
Dipole–induced dipole	Involves interactions between a permanent dipole (polar molecule) and a temporary dipole (typically a nonpolar molecule)	Ex: H_2O–CO_2	Weak
Ion–induced dipole	Involves interactions between an ion and a temporary dipole (typically a nonpolar molecule)	Ex: Ca^{2+}–O_2	Weak
Dipole–dipole	Involves permanent dipoles between polar molecules	Ex: CO–CO	Medium
Hydrogen bonding	A special type of dipole–dipole interaction between an H bound to an electronegative atom and an N, O, or F	Ex: HF–NH_3	Strong
Ion–dipole	Involves interactions between a polar molecule and an ion	Ex: H_2O–Na^+	Very strong

The type of IMF depends on the type of molecules involved in the interaction. Ions, polar molecules, and nonpolar molecules all experience different forces and different strengths of forces. For example, ions and polar molecules experience one of the stronger IMFs, ion–dipole. The weakest IMF (induced dipole–induced dipole) can be experienced by all molecules; however, it is the only force experienced between two nonpolar molecules. The polarizability of an electron cloud contributes to the strength of induced-dipole forces. *Polarizability* is the ability of an electron cloud to be distorted; larger atoms have bigger electron clouds and are more easily distorted. A more polarizable electron cloud experiences stronger induced-dipole forces because another molecule can more easily induce a dipole on it. Larger atoms or higher molecular weight molecules are more polarizable. Some molecules can also experience more than one IMF; for example, carbon monoxide (CO) can experience both dipole–dipole and London dispersion forces.

1. Distinguish induced-dipole forces from dipole forces. How are they alike? How are they different?

Deciding which IMFs are involved between two molecules can sometimes be confusing. It is crucial to be able to identify types of molecules: polar versus nonpolar. The decision chart in **Table 19.2** should help simplify these decisions.

TABLE 19.2 Types of Intermolecular Forces Between Molecules

Type of Interaction	Atoms	Nonpolar Molecules	Ions and Nonpolar Molecules	Nonpolar and Polar Molecules	Polar Molecules Without H, N, O, or F	Polar Molecules with H, N, O, or F	Ions and Polar Molecules
London dispersion	✓	✓	✓	✓	✓	✓	✓
Ion-induced dipole			✓				
Dipole–induced dipole				✓			
Dipole–dipole					✓	✓	
Hydrogen bonding						✓	
Ion–dipole							✓

2. How do boiling point and freezing point relate to IMFs in a liquid? Which IMFs would you suppose are associated more with solids than with liquids?

3. Identify the IMFs experienced by

   (a) NH_3

   (b) CH_4

   (c) CH_3Cl

   (d) CO_2

IMFs affect the properties of a liquid. Liquids composed of molecules that experience strong IMFs will have different properties than liquids composed of molecules that experience weak IMFs. Properties like boiling point, melting point, and vapor pressure are affected by the strength of IMFs in a liquid.

When classifying IMFs in liquids, we have two terms: cohesive forces and adhesive forces. *Adhesive forces* are IMFs between two different molecules, whereas *cohesive forces* represent the IMFs between identical molecules. Two cohesive forces that contribute to a liquid's properties are viscosity and surface tension.

Viscosity is the measure of a liquid's resistance to flow. Honey and oil are two examples of viscous liquids. Sugar molecules in honey experience hydrogen bonding, whereas the fatty acids in oil experience London dispersion forces.

Surface tension is the energy required to increase the surface area of a liquid by a given amount. Liquid water has a high surface tension, due to the structure created by the hydrogen bonding; some insects take advantage of this phenomenon and walk on water. You may have also noticed surface tension by the way water "beads up" on certain surfaces.

A prominent adhesive force is *capillary action*, which is the flow of liquid within a porous material due to its attraction to that material. You might have experienced this phenomenon when drying up water with a paper towel. Adhesive forces between water and glass are strong; hence, water is more attracted to the glass than to itself. This creates a meniscus in graduated cylinders. In the case of water and glass, the adhesive forces between the water and glass are stronger than the cohesive forces between the water molecules.

4. Oil and water form an immiscible mixture. From the perspective of water, indicate whether cohesive or adhesive forces are stronger. Explain.

By cooling, adding pressure, or both, liquids become solids and adopt additional properties. They can be hard, like a diamond, or soft, like butter. There are two main types of solids: amorphous and crystalline. A diamond is an example of a crystalline solid; butter is an example of an amorphous solid. Most solids are crystalline solids; however, let us examine the structure and properties of both.

Crystalline solids have particles that are arranged in an ordered manner, whereas amorphous solids have particles arranged randomly. The particles in crystalline solids all experience the same attractive forces, whereas the particles in amorphous solids experience a range of intermolecular forces, both type and strength. This leads to crystalline solids having a defined melting point at one temperature and amorphous solids melting over a range of temperatures. **Table 19.3** presents four types of crystalline solids.

TABLE 19.3 Crystalline Solids

Type of Crystalline Solid	Type of Particles	Type of Attractions	Properties	Examples
Ionic	Ions	Ionic bonds	Hard, brittle, conducts electricity but only as a liquid, poor thermal conduction, high melting point	$CaCl_2$, Fe_2O_3
Metallic	Atoms of electropositive elements	Metallic bonds	Malleable, ductile, excellent heat and electrical conduction, variable hardness and melting temperature	W, Ni, Fe, Pb, U
Covalent network	Atoms of electronegative elements	Covalent bonds	Very hard, not conductive, very high melting point, variable thermal and electrical conduction	C (diamond), SiO_2, SiC, quartz
Molecular	Molecules	IMFs	Fairly soft, variable brittleness, poor conduction of heat or electricity, low melting point	H_2O, CH_4, CO_2, I_2, $C_{12}H_{22}O_{11}$

5. Using spheres, draw at the molecular level a picture of a crystalline solid and a picture of an amorphous solid.

6. Ethanol melts at −173.5 °C, whereas atactic polystyrene (packing peanuts) melts between 100.0 and 120.0 °C. Explain the structure of these solids at the molecular level. Do you think the IMFs in the polystyrene solid have similar or different energies than those in ethanol? Why?

The structure of crystalline solids is described using a unit cell. The *unit cell* is the smallest repeating unit of a crystalline structure. Let's examine the structural possibilities with a cubic unit cell; there are many other lattice systems with different unit cell shapes, but we will focus on the cubic. A *cubic unit cell* has sides of equal length that intersect at 90° angles; it is an imaginary cube that we use to describe a crystalline solid structure. Many metallic compounds form solids with these cubic unit cell structures. The atoms in a cubic unit cell are represented by hard spheres packed next to and on top of each other at lattice points. The different ways these hard spheres can be stacked in the unit cell produce the different types of unit cells. **Table 19.4** highlights the three types of cubic unit cells.

Looking at the chart, let's discuss a simple cubic unit cell. A simple cubic cell has 1 atom inside of it because there is ⅛ of each atom at the 8 corners of the unit cell. Each atom in a simple cubic touches 6 other atoms; therefore, it has a coordination number of 6. A simple cubic is the least efficient in terms of space; atoms occupy only 52% of the space inside the unit cell. Determining the length of a simple cubic is just twice the atomic radius because the atoms lie on the edge of the unit cell. Using Table 19.4, note the space efficiencies and coordination numbers of body-centered cubic and face-centered cubic unit cells.

TABLE 19.4 Cubic Unit Cells

Type of Cubic Unit Cell	Representation	Number of Atoms	Coordination Number	Packing Efficiency	Length of Side (*a*)
Simple		1	6	52%	$a = 2r$
Body-centered		2	8	68%	$a = \dfrac{4}{\sqrt{3}}r$
Face-centered		4	12	74%	$a = r\sqrt{?}$

Ionic compounds consist of two or more atoms; therefore, the unit cells of ionic compounds sometimes differ from the unit cells of metallic structures. An ionic compound with a 1:1 ratio and similar cation and anion radii tends to form a simple cubic unit cell, whereas an ionic compound with a 1:1 ratio and a significant difference in ionic radii tends to form a body-centered cubic. Ionic compounds forming face-centered cubic unit cells are not as easily distinguished.

7. Why might an ionic compound, such as NaCl, adopt a body-centered cubic?

Review Question

8. Wax is composed of long-chain, high molecular weight, nonpolar hydro-carbons, whereas water is composed of a polar, small molecule. Write a paragraph discussing the relationship of IMFs and the properties of these substances. Make sure to highlight their melting and boiling points, molecular weight, viscosity, polarity, and so on.

Application Questions

9. Arrange the following sets of compounds by decreasing IMF strength.

   (a) HF

   (b) CH_4

   (c) CO_2

   (d) He

   (e) NO

10. Arrange the following sets of compounds by increasing boiling point.

    (a) C_3H_8

    (b) H_2O

    (c) N_2

    (d) SiH_4

    (e) CH_2Cl_2

Use the chart to answer the following question.

	Boiling Point (°C)	**Surface Tension (J/m²)**
Ethanol, CH_3CH_2OH	78	2.3×10^{-2}
Propanol, $CH_3CH_2CH_2OH$	97	2.4×10^{-2}
n-Butanol, $CH_3CH_2CH_2CH_2OH$	117	2.6×10^{-2}

11. The molecules in the chart all experience the same intermolecular forces. If they experience the same intermolecular forces, explain the trends in boiling points and surface tensions.

12. Indicate the type of crystalline solid for the following substances.

   (a) SiO_2

   (b) Au

   (c) $CaCl_2$

   (d) $C_4H_{10}O$

   (e) I_2

13. You are tasked with identifying an unknown substance. It melts at 825 °C, is partially soluble in water, and conducts electricity in water. The appearance is hard and brittle. What type of crystalline solid do you have (ionic, metallic, covalent network, or molecular)?

14. How many atoms are there in a unit cell of tungsten, which is crystallized in a body-centered cubic structure? What is the coordination number of each tungsten atom?

15. Determine the metallic radius of a silver atom from X-ray diffraction data that revealed that the structure is face-centered cubic with a unit cell length of 466.7 pm.

Figure Credits

CHAPTER 20

Solutions

What are the different ways to measure concentration?

Are all solutions just compounds dissolved in water?

Learning Outcomes

- Explain how solutions are formed.
- Demonstrate the ability to calculate solutions in varying concentration units, such as molarity, molality, w/w, v/v, w/v, and ppm.

Prerequisite Knowledge

- Chapter 6: Chemical Bonding
- Chapter 15: Limiting Reactants and Molarity

Background Information and Questions

Solutions are a core concept in chemistry. They are a mixture of two or more compounds in a single phase (solid, liquid, or gas). The compound of which there is the largest amount is the *solvent*, and the compounds that are dissolved in it are the *solutes*. The majority of solutions that you will encounter in this book are aqueous solutions, which are represented by an (aq) and use water as the solvent.

A standard rule of solutions is "like dissolves like," which refers to the polarity of the solute and solvent. Polar solutes dissolve more readily in polar solvents and vice versa.

1. Think back to the chapter on chemical bonding (Chapter 6). Which forces held polar molecules together?

2. Which forces held nonpolar molecules together?

3. How do you think the rule "like dissolves like" works in terms of intermolecular forces? *Hint: Think about how polar molecules interact with each other!*

Ionic compounds also dissolve readily in polar solvents. When an ionic compound dissolves, the cations are surrounded by the negative dipoles of the solvent, while the anions are surrounded by the positive dipoles.

4. NaCl is an ionic compound that readily dissolves in water. Sketch what it would look like when the cation (sodium) is surrounded by water molecules as described in the paragraph above.

5. NaCl is an ionic compound that readily dissolves in water. Sketch what it would look like when the anion (chlorine) is surrounded by water molecules as described in the paragraph above.

6. Will the following compounds dissolve more readily in H_2O or C_8H_{18}? Why?

(a) NaCl

(b) SO_2

(c) CO_2

Not all solutions are aqueous or even in liquid form, however. Some alloys are solutions in which one metal is dissolved in another to produce a solid solution. An example of a gaseous solution would be the air we breathe, in which gases such as oxygen and hydrogen are dissolved in nitrogen.

The formation of any solution is a spontaneous process. Motion (such as stirring) or heat can speed the process of dissolution, but it will eventually occur regardless. There are three steps in the formation of a solution: (a) solute particles are separated from each other, (b) solvent particles are separated from each other, and (c) the solute and solvent particles combine to form the solution. Steps 1 and 2 are endothermic and require energy, but step 3 is exothermic and releases energy.

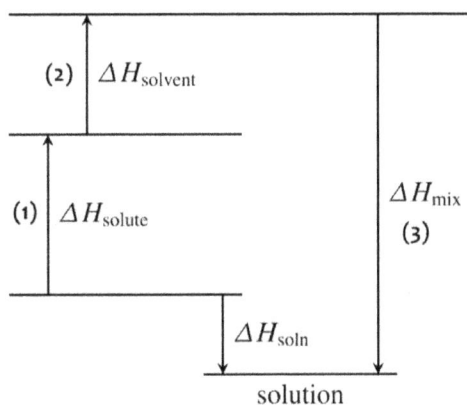

FIGURE 20.1 Enthalpies associated with solution formation.

Concentration is a measure of the amount of solute in the amount of solution (or solvent) present. The most common unit of concentration is molarity, which is calculated as moles of solute per liter of solution (see Chapter 15).

7. *Review!* 5.0 moles of NaCl are dissolved in 2.0 L of water. What is the molarity?

Normality is a unit most commonly found when working with acids and bases. Instead of using moles, normality instead defines concentration by using "equivalents." An equivalent is the amount of a chemical species reacting stoichiometrically with another chemical species. *This means that normality is a function of the chemical reaction that is taking place.*

A simple example of normality is H_2SO_4. There is a relationship between normality and molarity. Normality can only be calculated when we deal with reactions because normality is a function of equivalents. Normality refers to compounds that have multiple chemical functionalities, such as sulfuric acid, H_2SO_4. A 1 M solution of H_2SO_4 will contain only 1 mole of H_2SO_4 in 1 liter of solution, but if the solution is titrated with a base, it will be shown to contain two moles of acid. This is because a single molecule of H_2SO_4 contains two acidic protons (H^+ ions). Thus, a 1 M solution of H_2SO_4 will be 2 N. The normality of a solution is the molarity multiplied by the number of equivalents per mole.

8. What would the normality (N) of a 1.0 M solution of HCl be?

Generally used in thermodynamic calculations that require a temperature-independent unit for concentration, molality uses the solvent's mass (in kilograms) instead of the solution's volume to make a unit that is not temperature dependent. The most commonly used units are moles of solute per kilogram of solvent.

$$m = \frac{\text{mol solute}}{\text{kg solvent}}$$

What is the molality of a solution when 0.65 moles of solute are dissolved in 2.0 kg of solvent?

$$m = \frac{0.65 \text{ mol}}{2.0 \text{ kg}} = 0.325 \text{ m}$$

What is the molality of a solution when 5.0 moles of NaCl are dissolved in 20.0 L of water? Recall that the density of water is approximately 1 g/mL; therefore, 20.0 L would weigh approximately 20.0 kg (this trick only works with water!).

$$m = \frac{5.0 \text{ mol NaCl}}{20.0 \text{ kg H}_2\text{O}} = 0.25 \text{ m}$$

9. 5.0 moles of NaCl are dissolved in 2.0 kg of water. What is the molality?

This text primarily focuses on molarity as a concentration, with the occasional use of molality. However, it is important to be familiar with other forms of concentration as well. Some of these are as follows:

$$\text{Weight percent (w/w)} = \frac{\text{weight solute}}{\text{weight solution}} \times 100\%$$

$$\text{Volume percent (v/v)} = \frac{\text{volume solute}}{\text{volume solution}} \times 100\%$$

$$\text{weight/volume percent (w/v)} = \frac{\text{weight solute, g}}{\text{volume solution, mL}} \times 100\%$$

Weight percent is often used to express the concentration of commercial aqueous reagents. For example, nitric acid is sold as a 70% (w/w) solution, meaning that the reagent contains 70 g of HNO_3 per 100 g of solution.

Volume percent is commonly used to specify the concentration of a solution prepared by diluting a pure liquid compound with another liquid. For example, a 5% (v/v) aqueous solution of methanol *usually* describes a solution prepared by diluting 5.0 mL of pure methanol with enough water to give 100 mL, which would be 95 mL of H_2O.

Weight/volume percent is often used to indicate the composition of dilute aqueous solutions of solid reagents. For example, 5% (w/v) aqueous silver nitrate *often* refers to a solution prepared by dissolving 5 g of silver nitrate in sufficient water to give 100 mL of solution.

For very dilute solutions, parts per million (ppm) and parts per billion (ppb) are convenient. On another note, it is possible to think of "percent" as being parts per hundred (pph).

$$\text{ppm} = \frac{\text{mass solute}}{\text{mass solution}} \times 10^6 = \frac{\text{mass solute (mg)}}{\text{volume solution (L)}}$$

Parts per billion can be calculated in a similar manner but with 10^9 instead of 10^6. Parts per million can be estimated as milligrams per liter (mg/L) in solutions that have a density of near 1.00 g/mL. *Most* solutions that are dilute enough to warrant using parts per million will work with this trick.

10. A solution contains 50.0 mg of solute in 2.00 L of solution. What is the concentration in ppm?

Review Questions

11. Why are the first two steps in dissolving a solute into a solvent endothermic? Why is the third step exothermic? Explain.

12. When is normality most commonly used as a unit of concentration?

13. A 50.0% w/w solution of NaOH would contain how many grams of NaOH per every 100.0 g of solution?

14. Why can we assume that ppm is the same as mg/L when the density of a solution is close to 1 g/mL?

15. Why is molality a better choice than molarity when temperature changes are involved?

Application Questions

16. What is the molality of a solution that contains 44.0 g NaCl in 500.0 g of water?

17. If 30.0 g of lithium chloride is dissolved in 200.0 g of water, what is the weight percent of lithium chloride in the solution?

18. Calculate the molarity for the solutions described in the two previous questions.

Figure Credit

CHAPTER 21

Colligative Properties

Why do we add salt to roads in the winter?

Why does your car have antifreeze?

How does osmosis work to regulate water in cells?

Why is reverse osmosis used to purify water?

Learning Goals

- Identify and explain each colligative property: vapor pressure, freezing point, boiling point, and osmotic pressure.
- Calculate the molar mass of an unknown solute based on information from the colligative property.
- Determine the osmotic pressure and direction of flow for a biological cell.

Prerequisite Knowledge

- Chapter 19: Liquids and Solids
- Chapter 20: Solutions

Background Information and Questions

Colligative properties are used to explain the behavior of solutions. They depend on asking "how much" instead of "what" when it comes to solutes; that is, the colligative properties of a solution depend on the number of nonvolatile solute particles in the solution rather than the type or identity of the solute particles that are present. These properties include vapor pressure lowering, boiling point elevation, freezing point depression, and osmotic pressure.

Because the colligative properties of a solution depend on how many particles are present, one must consider what happens to the solute once it is in the solution. A good way to illustrate this concept is to compare the effects of sugar versus salt in water. Recall that, as discussed in Chapter 20, salt is an electrolyte that dissociates in water, and sugar is a nonelectrolyte.

Sucrose ($C_{12}H_{22}O_{11}$) is commonly known as table sugar. Because sucrose is nonionic, it will remain in the form $C_{12}H_{22}O_{11}$ in solution rather than splitting into ions when it is dissolved in the solvent (water). What about salt? Table salt (NaCl) is an ionic compound that will separate into the ions Na^+ and Cl^- when dissolved in water (see **Figure 21.1**).

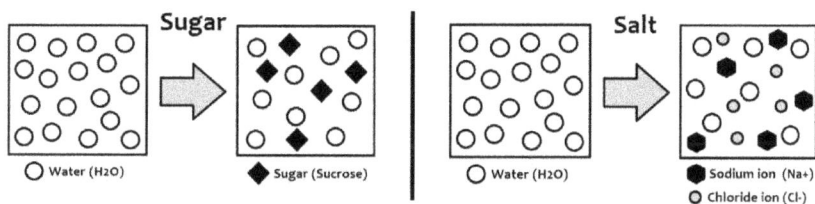

FIGURE 21.1 Sugar versus salt in solution.

We get *twice* as many particles in solution when we add salt instead of sugar! Therefore, the salt will have a greater effect on the colligative properties of a solution than sugar. This ratio between the actual concentration of particles in solution and the concentration of the solute is called the *van't Hoff factor* (*i*). We will estimate the van't Hoff factor as the number of particles generated when a solute goes into solution (for NaCl $i \approx 2$, but for $C_{12}H_{22}O_{11}$ $i = 1$ because sucrose does not dissociate).

> Note: The actual value for the van't Hoff factor in a solution containing an ionic solid is slightly lower than our estimates due to ion pairing. Ion pairing is a concept that accounts for the fact that occasionally a cation (such as Na^+) and an anion (such as Cl^-) may come together to form a single ion pair in solution.

1. Assume you have 1.0 mole of each of the following compounds. Put them in order from most to least effect on colligative properties. *Hint! Remember to check what each compound would look like in solution in terms of ions and determine which would produce the most particles.*

    KI $C_6H_{12}O_6$ Na_3PO_4 $MgCl_2$

As for the colligative properties themselves, let's begin with the *lowering of vapor pressure*. Vapor pressure is the pressure exerted by the vapor molecules above a liquid. Recall that some volatile (easily vaporized) liquids have molecules that are in the vapor phase above the liquid's surface. If there is a nonvolatile solute dissolved in the solvent, there are fewer volatile solvent molecules at the surface. Fewer solvent molecules at the surface means fewer of them evaporate; therefore, the vapor pressure is lowered. We can calculate the vapor pressure associated with a solution containing a nonvolatile solute using Raoult's law:

$$P_{\text{solution}} = X_{\text{solvent}} P^0_{\text{solvent}}$$

In this equation, P_{solution} will be the vapor pressure of the solution, P^0_{solvent} the vapor pressure of the pure solvent, and X_{solvent} the mole fraction of the solvent in the solution. But what if the solute is *also* volatile? In that case, both the solvent and the solute(s) will be exerting a vapor pressure that must be taken into account. For this situation, we can modify Raoult's law to account for the partial pressures of the vapor mixture:

$$P_{\text{solution}} = X_A P^0_A + X_B P^0_B + \ldots$$

2. If you added 0.20 mol NaCl to 5.0 mol H_2O, what would the new vapor pressure be for the salt water? *Hint! Remember that mole fraction is moles of a compound over the total number of moles.* Info: $T = 25.0°C$, $P^0_{H_2O} = 23.8$ mmHg

The boiling point of a solution is elevated above the boiling point of a pure solvent. Because we know that a liquid boils once its vapor pressure reaches atmospheric pressure, the concept of *boiling point elevation* can be explained using the lowering of vapor pressure. If the vapor pressure is lower than that of the pure solvent, a higher temperature will be required for the solution's vapor pressure to reach atmospheric pressure or boil. This increase in boiling point is directly proportional to the concentration of the solute:

$$\Delta T = iK_b m$$

In this equation ΔT is the change in temperature of the boiling point, i is the van't Hoff factor, K_b is a constant unique to each solvent, and m is the molality, which is moles of solute per kilogram of solvent. An analogous equation using K_f instead of K_b is available for calculating the *freezing point depression:*

$$\Delta T = iK_f m$$

Note: Some textbooks present the K_f value as negative. This assumes that $\Delta T = T_{final} - T_{initial}$ rather than the more common $\Delta T = T_{solvent} - T_{solution}$. We will present them as positive values, like the majority of textbooks. You will need to subtract the change in freezing point from the freezing point of the solvent and add the change in boiling point to the boiling point of the pure solvent.

The boiling point elevation and freezing point depression can also be explained by thinking of solute–solvent interactions and solvent–solvent interactions. Solute–solvent interactions in a solution are stronger than solvent–solvent interactions in a pure solvent; therefore, it takes more energy to overcome the solute–solvent interactions in a solution than a pure solvent. Hence, the boiling point is elevated for the solution. The freezing point is depressed because the solvent–solvent interactions are weaker than solute–solvent interactions; therefore, the solution has to be cooled more in order for the solvent–solvent interactions to prevail and form the solid.

3. Why is molality used instead of molarity when calculating freezing point depression and boiling point elevation? (Stop and *think* about this; you can figure it out!)

4. If 0.20 mol NaCl is added to 1.50 kg H_2O, what is the new freezing point of the water? Info: K_f for H_2O = 1.86 °C/m

The final colligative property we need to understand is that of *osmotic pressure*. Osmosis is the movement of a solvent through a semipermeable membrane from an area of lower solute concentration to an area of higher solute concentration. In other words, osmosis is diluting the more concentrated solution, but since the sol*ute* particles can't move through the semipermeable membrane, it is the solvent that must move instead of the sol*ute*. This leads to a difference in concentration of solute across the membrane.

The osmotic pressure of a solution is the difference in pressure required to stop the flow of *solvent*. It is given by the following equation:

$$\Pi = iMRT$$

In this equation, Π is the osmotic pressure, i is the van't Hoff factor, M is the molarity, R is the ideal gas constant, and T is the temperature.

Tonicity is commonly used to discuss osmotic pressure. In **Figure 21.2**, the arrows indicate the direction that the *solvent* is moving. For example, a red blood cell placed into a hypotonic solution (less saline than your body's cells) would swell (or even burst) as water moves into the cell in an attempt to equalize the osmotic pressure. Conversely, a red blood cell placed into a hypertonic solution (more saline than your body's cells) would shrivel as water moves out. Red blood cells in the body are typically in isotonic solution (the same concentration as your body's cells).

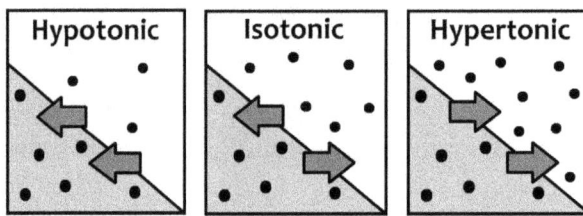

FIGURE 21.2 Tonicity.

5. Assuming that red blood cells have an NaCl concentration the same as that of saline solutions (9.0 g NaCl per liter H_2O), what would happen if you placed red blood cells into

   (a) saline solution

   (b) seawater (35.0 g NaCl per liter H_2O)

   (c) distilled water

Review Questions

6. Pickles are made when cucumbers are placed in brine, which is a solution with a high salt content. Explain the pickling process in terms of osmosis.

7. List the four colligative properties and give brief descriptions of their effects on solutions.

8. Why is the magnitude of colligative properties usually proportional to the concentration of the solute?

9. Recall the relationship of vapor pressure and boiling point. Explain this relationship in terms of colligative properties.

Application Questions

10. Assume you add 50.0 g of each of the following compounds to 500.0 mL of water. Put them in order from *most* to *least* effect on colligative properties.

KI	$C_6H_{12}O_6$	Na_3PO_4	$MgCl_2$
166.0 g/mol	180.2 g/mol	163.9 g/mol	95.21 g/mol

11. It is common to add a pinch of salt to water when cooking. What would be the new vapor pressure and boiling point of 1.500 L of H_2O if 16.000 g of NaCl were added (roughly ½ tablespoon)? Info: $T = 25\,°C$, $P^0_{H_2O} = 23.8$ mmHg, K_b for $H_2O = 0.512\,°C/m$

12. Antifreeze has several uses, one of which is to prevent water from freezing, expanding, and bursting its container. One of the most common antifreeze solutions is composed primarily of ethylene glycol ($C_2H_6O_2$, 62.07 g/mol). If 20.00 g of antifreeze were mixed with 50.00 mL of water, what would the new freezing point be? Info: K_f for $H_2O = 1.86\,°C/m$

13. The vapor pressure of hexane (C_6H_{14}, 86.17 g/mol) is 124.0 mmHg at 25 °C. How much is the vapor pressure lowered if 3.60 g of warfarin ($C_{19}H_{16}O_4$, 308.3 g/mol), a blood thinner, is added to 128.0 g of hexane?

14. Vodka does not freeze when placed into a standard freezer, due to the high ethanol content (CH_3CH_2OH, 46.07 g/mol). The ethanol lowers the freezing point of the water! But as a volatile compound, the ethanol also has a strong effect on vapor pressure. If you have a mixed drink that contains approximately 50.0 g ethanol and 50.0 g water, what would be the vapor pressure of this solution? Info: $T = 0$ °C, $P^0_{H_2O} = 4.6$ mmHg, and $P^0_{EtOH} = 12$ mmHg

15. A normal saline solution contains 9.0 g of NaCl (58.44 g/mol) per liter of water (0.9% w/v), while some seawater samples contain 35.0 g of salts per liter. Assuming that the seawater sample only contains NaCl (it doesn't), calculate the osmotic pressure of each of these solutions at 25 °C. Info: $R = 0.08206 \dfrac{L*atm}{mol*K}$

16. Xylitol is an artificial sweetener made from birch tree bark. Calculate the molecular weight of xylitol if 4.50 g of it is dissolved in 250.0 g of ethanol (C_2H_6O, 46.07 g/mol). Info: K_b = 1.22 °C/m, ΔT_b = 0.14 °C

Kinetics Concepts

How quickly are reactants converted into products?

How do different experimental values affect this rate?

How can we describe and study reaction rates?

Learning Outcomes

▧ Describe a reaction in terms of the change in concentration per unit time.

▧ Describe a reaction in the form of a differential rate law and explain how the rate depends on concentration.

▧ Determine a rate law using experimental data and the method of initial rates.

▧ Identify the overall order of reaction.

Prerequisite Knowledge

▧ Chapter 15: Limiting Reactants and Molarity

Background Information and Questions

Chemical kinetics is the study of the rate (or speed) at which reactants are converted into products in a chemical reaction. This rate depends on several factors, such as the physical state, temperature, and concentration of the reactants.

The *rate of reaction* is the change in the amount of reactant or product per unit time. Most often this is expressed in terms of concentration, as shown below for a reaction in which a reactant, NO_2, decomposes:

$$NO_2 \text{ (g)} \rightarrow NO \text{ (g)} + \frac{1}{2} O_2 \text{ (g)}$$

$$\text{Rate of decomposition of } NO_2 = -\frac{[NO_2]_{Final} - [NO_2]_{Initial}}{\text{Amount of time passed}}$$

$$\text{Rate of decomposition of } NO_2 = -\frac{\Delta[NO_2]}{\Delta t}$$

The rate equation is written with a negative sign because rates are listed as positive numbers, and in this case the NO_2 (reactant) is disappearing (the final amount present will be less than the initial amount). The negative sign is required to "cancel out" this additional negative quantity. The rate expression for the product would have a positive sign since it is the appearance, not the disappearance, that is being monitored:

$$\text{Rate of formation of } NO = \frac{\Delta[NO]}{\Delta t}$$

Water can be broken down to give hydrogen and oxygen:

$$2\ H_2O \rightarrow 2\ H_2 + O_2$$

1. Write an equation for the rate of decomposition of water.

2. Write an equation for the rate of appearance of hydrogen.

3. Write an equation for the rate of appearance of oxygen.

4. If this reaction started with 2.0 M H_2O_2 and 30 seconds later the concentration of H_2O_2 was 0.5 M, what would be the rate of decomposition? *Hint: Plug in the final concentration (0.5 M) minus the initial concentration (2.0 M) for the numerator, then plug in 30 s for the change in time in the denominator. H_2O_2 is a reactant, so don't forget the negative sign!*

If it is not possible to directly measure the rate of reaction, a *relative* rate of reaction can be determined from the balanced chemical equation. Consider again the decomposition of water:

$$2\,H_2O \rightarrow 2\,H_2 + O_2$$

The stoichiometry of the equation can be used to relate reaction rates and determine the rate of each chemical in comparison to the others. For every water molecule that decomposes, a hydrogen molecule is formed; it follows that the rates for those two chemicals should be the same because they have the same coefficients and are at 1:1 ratio in the equation. However, the hydrogen is appearing twice as quickly as the oxygen because for every two hydrogen molecules only one oxygen appears. Remember, the coefficient becomes the denominator when writing relative rates. Therefore, we can write the rates as follows in relation to each other:

$$-\frac{1}{2}\frac{\Delta[H_2O]}{\Delta t} = \frac{1}{2}\frac{\Delta[H_2]}{\Delta t} = \frac{\Delta[O_2]}{\Delta t}$$

This relationship of relative rates is shown graphically in **Figure 22.1**. The concentration of hydrogen is appearing at the same rate as water is disappearing, while oxygen only appears half as rapidly. Notice that as the reaction proceeds, the concentrations appear to change more slowly; that is, the reaction slows down. At equilibrium the concentrations will stabilize at their final value.

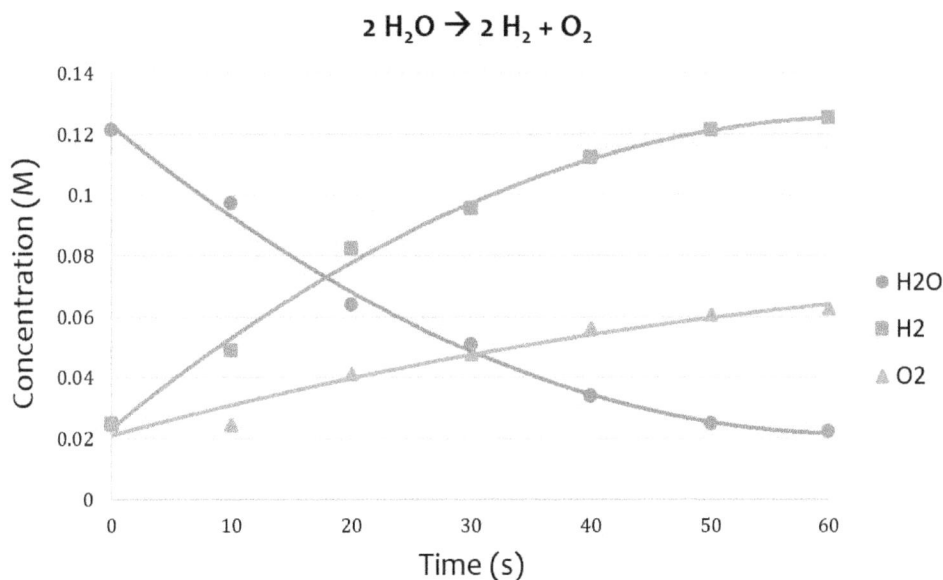

FIGURE 22.1 Relationship of relative rates.

5. Write the relative rates for all compounds in the following reaction:

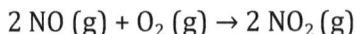

$$2\ NO\ (g) + O_2\ (g) \rightarrow 2\ NO_2\ (g)$$

A *rate law* relates the rate to the concentration of the reactants. In this chapter we will focus on the *differential rate law*, which expresses the rate in terms of changes of reactant concentration over a period of time. The following is a typical rate law:

$$\text{Rate} = k[A]^{\alpha}[B]^{\beta}$$

in which k is the *rate constant*, [A] and [B] are reactant concentrations, and the exponents indicate the *reaction order*. The rate constant is constant for a given reaction but can vary with temperature or other reaction conditions.

The reaction order indicates the extent to which the rate depends on each individual reactant. These exponents are typically whole number values of 0, 1, or 2, though this is not always the case. A higher order for a particular reactant means that the rate of the reaction is depending more on that reactant than on others. Changes in concentration of a zero-order reactant have no effect on the rate of the reaction, while changes in concentration of a first order reactant have a linear effect, and a second order reactant has an exponential efficient.

$$\text{Rate} = k[A][B]^2$$

The above reaction is said to be first order in A (the exponent is assumed to be 1), second order in B, and third order overall (add the exponents together to get the overall order of reaction).

Consider the following chemical reaction and its differential rate law:

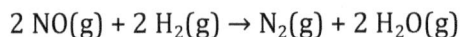

$$2\ NO(g) + 2\ H_2(g) \rightarrow N_2(g) + 2\ H_2O(g)$$
$$\text{Rate} = k[NO]^2[H_2]$$

6. What is the order of reaction with respect to NO?

7. What is the order of reaction with respect to H_2?

8. What is the overall order of reaction?

9. Changing the concentration of *which* reactant would have the greatest effect on rate?

Notice that the values for reaction order do not match the coefficients in the balanced chemical equation. Both the reaction order and the rate constant must be determined experimentally. This is accomplished using the *method of initial rates*. Consider the following example and the information in **Table 22.1**:

$$A + 2B + 3C \rightarrow 2D$$

TABLE 22.1 Method of Initial Rates

Trial	Initial Concentration [A]	Initial Concentration [B]	Initial Concentration [C]	Initial Rate of D (M/s)
1	0.2	0.3	0.1	4×10^{-3}
2	0.2	0.6	0.1	8×10^{-3}
3	0.4	0.6	0.1	3.2×10^{-2}
4	0.2	0.3	0.2	4×10^{-3}

To determine the order of reaction with respect to each reactant, we need to compare the change in concentration of one reactant (while others are held constant) to the change in the rate of reaction.

For [A]: Compare trials 2 and 3 because the concentration of [A] changes while the concentrations of [B] and [C] do not.

When [A] *doubles* from 0.2 M to 0.4 M, the rate *quadruples* from 0.008 M/s to 0.032 M/s. So when [A] went up by a factor of 2, the rate went up by 2^2. This exponent tells us that the order with respect to A will be 2.

For [B]: Compare trials 1 and 2 because the concentration of [B] changes while the other concentrations do not.

When [B] *doubles* from 0.3 M to 0.6 M, the rate also *doubles* from 0.004 M/s to 0.008 M/s. So when [B] went up by a factor of 2, the rate went up by 2^1. This exponent tells us that the order with respect to B will be 1.

For [C]: Compare trials 1 and 4 because the concentration of [C] changes while the other concentrations do not.

When [C] *doubles* from 0.1 to 0.2, the rate *doesn't change* at all. So when [C] went up by a factor of 2, the rate went up by 2^0. This exponent tells us that the order with respect to C will be 0.

Combining those values into a rate law gives us the following equation with an overall reaction of 3. The equation should be rewritten to denote that a reactant with an exponent of 0 does not affect the rate.

$$\text{Rate} = k[A]^2[B]^1[C]^0$$

$$\text{Rate} = k[A]^2[B]$$

To get the rate constant, we can pick values from the table, plug in, and solve for k. The following utilizes trial 1 for this purpose:

$$0.004 \text{ M/s} = k(0.2 \text{ M})^2(0.3 \text{ M})$$

$$k = 0.33 \text{ M}^{-2}\text{s}^{-1}$$

The rate law for this chemical reaction under these conditions is therefore:

$$\text{Rate} = 0.33 \text{ M}^{-2}\text{s}^{-1} [A]^2[B]$$

Use the information provided to answer the following questions:

$$2A + 2B \rightarrow C + 2D$$

Trial	Initial Concentration [A]	Initial Concentration [B]	Initial Rate (M/s)
1	0.4	0.3	2×10^{-3}
2	0.4	0.6	4×10^{-3}
3	0.8	0.3	8×10^{-3}

10. What is the order of reactant A?

11. Is reactant B zero order, first order, second order, or third order?

12. What is the overall rate of reaction?

13. What is the rate constant?

14. What are the units of the rate constant? *Hint: Depending on the overall order of reaction, this will change!*

15. Write the differential rate law for this reaction under these conditions.

Review Questions

16. What would the units for *k* be if the reaction was zeroth, first, second, and third order? *Hint: These will be different values!*

17. The overall order of reaction determines the units for *k*; why is that?

Application Questions

Rate data was obtained for the reaction $A + B \rightarrow C$.

Trial	Initial Concentration [A]	Initial Concentration [B]	Initial Rate (mol/L*s)
1	3	4	0.4
2	6	4	0.8
3	6	8	0.8

18. What is the order of reactant A?

19. What is the order of reactant B?

20. What is the overall order of this reaction?

21. Write the differential rate equation for this reaction.

22. What is the rate constant for this reaction?

Rate data was obtained for the reaction A + 2B + C → 3D.

Trial	[A]	[B]	[C]	Rate
1	1.0×10^{-3}	0.45	1.0×10^{-3}	3.50×10^{-1}
2	2.0×10^{-3}	0.45	1.0×10^{-3}	3.50×10^{-1}
3	1.0×10^{-3}	1.80	1.0×10^{-3}	5.60×10^{-0}
4	1.0×10^{-3}	0.45	3.0×10^{-3}	1.05×10^{-0}

23. Write the rate law for this chemical reaction.

24. If you wanted to slow down this reaction to better study it, which reactant (A, B, or C) concentration would you decrease in order to have the greatest impact on rate?

Consider the reaction between nitrogen monoxide and chlorine:

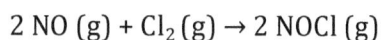

$$2\ NO\ (g) + Cl_2\ (g) \rightarrow 2\ NOCl\ (g)$$

Trial	Initial Concentration [NO]	Initial Concentration [Cl_2]	Initial Rate NOCl (mol/L*s)
1	0.006	0.006	0.05
2	0.012	0.006	0.20
3	0.006	0.012	0.10

25. What is the rate constant, k, for this reaction?

26. Predict the rate of this reaction if the initial concentrations were [NO] = 0.50 M and [Cl$_2$] = 0.50 M.

CHAPTER 23

Integrated Rate Laws

How long does it take a reaction to react?

What is half-life?

Learning Outcomes

- Use integrated rate laws to determine the rate constant.
- Describe the half-life of a chemical reaction and calculate its value.
- Determine whether each half-life equation is zeroth, first, or second order.

Prerequisite Knowledge

- Chapter 22: Kinetics Concepts

Background Information and Questions

In the previous chapter, we discussed the concept of rate and the differential rate laws, which study rate as a function of concentration. The integrated rate laws are used to study rate as a function of time. For example, an integrated rate law can estimate the amount of time required for a reaction to proceed to a certain point. Integrated rate laws are derived by using calculus to integrate the differential rate laws with respect to time and are as follows:

$$Zeroth-order\ reation: [A] = -kt + [A]_0$$

$$First-order\ reaction: ln[A] = -kt + ln[A]_0$$

$$Second-order\ reaction: \frac{1}{[A]} = kt + \frac{1}{[A]_0}$$

$$Straight\ line\ equation: y = mx + b$$

in which [A] is the concentration of a reactant at a specific time, $[A]_0$ is the amount of initial reactant present, t is the time that has passed, and k is the rate constant. All three of these equations can be graphed if we map them onto the equation for a straight line, $y = mx + b$. The graphs are shown in **Figure 23.1**.

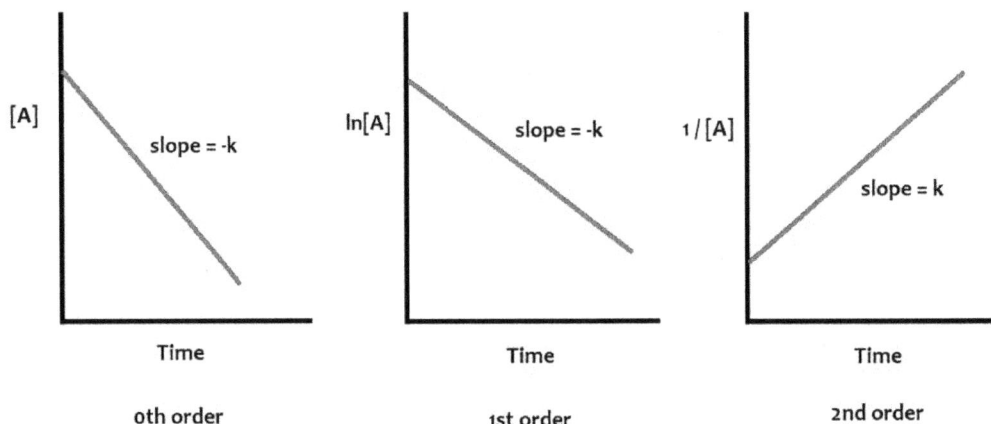

FIGURE 23.1 Graphs of order of reaction.

In each case the value for the y-axis changes with the order of reaction (see Figure 23.1). If graphed simply as concentration versus time, first- and second-order reactions would not produce a straight line, and therefore, the reaction constant k (the slope) could not be determined. To produce a straight line, first-order integrated rate laws must be plotted as ln[A] versus time, while second-order integrated rate laws must be graphed as 1/[A] versus time. Figure 23.1 illustrates the changes in the y-axis based on the order of the integrated rate law.

1. Kinetic data is obtained for the decomposition of dinitrogen pentoxide, N_2O_5. When graphed as [A] versus t, a curved line appears. When graphed as ln[A] versus t, a straight line appears. When graphed as 1/[A] versus t, a curved line appears.

    (a) What is the order of this reaction?

(b) If the slope of the straight line is -4.5×10^{-4}, what is the rate constant for this reaction?

The integrated rate laws can also be used to determine how long it will take for a reaction to proceed to a certain point. For example, the rate constant for the following reaction is $0.008 \ s^{-1}$:

$$A + B \rightarrow C$$

If this is known to be a first-order reaction, how long will it take for 2.0 M of [A] to be reduced to 0.5 M [A] (producing 1.5 M [C] in the process)? Use the integrated rate law for a first-order reaction!

$$ln\,[A] = -kt + ln\,[A]_0$$

$$ln\,(0.5M) = -(0.008\ s^{-1})\,(t) + ln\,(2.0\ M)$$

$$t = 173\ s$$

2. The rate constant of the following first-order reaction is known to be $0.15 \ s^{-1}$:

$$A \rightarrow B$$

How long will it take for this reaction to reduce an initial concentration of 4.0 M [A] to 0.6 M [A]?

The *half-life* of a reaction is the time required for one half of a given amount of reactant to be used up. Half-life is given as $t_{1/2}$ and is most often encountered in nuclear reactions, though it is also important for pharmacology kinetics and other areas of chemistry. For a first-order reaction, half-life can be illustrated via a chart, such as the one shown in **Table 23.1** for a reaction starting with 5.0 M reactant concentration.

TABLE 23.1 Example of First-Order Reaction Half-Life

Number of Half-Lives Passed	Percentage Reactant Left	Amount Reactant Left (M)	Fraction Remaining
0	100.000%	5.00	1
1	50.000%	2.50	$\frac{1}{2}$
2	25.000%	1.25	$\frac{1}{4}$
3	12.500%	0.63	$\frac{1}{8}$
4	6.250%	0.31	$\frac{1}{16}$
5	3.125%	0.16	$\frac{1}{32}$
6	1.563%	0.08	$\frac{1}{64}$
7	0.781%	0.04	$\frac{1}{128}$
8	0.391%	0.02	$\frac{1}{256}$

Note that with each half-life, the amount of reactant is halved. If the half-life for Table 23.1 was 1 hour, then half of the reactant would disappear for each hour that passed. Equations for half-life can be obtained from the integrated rate laws by subbing in ½[A]₀ for the value of [A] and solving for *t*. These half-life equations are as follows:

$$Zeroth-order\ half-life:\ t_{1/2} = \frac{[A]_0}{2k}$$

$$First-order\ half-life:\ t_{1/2} = \frac{0.693}{k}$$

$$Second-order\ half-life:\ t_{1/2} = \frac{1}{k[A]_0}$$

The majority of reactions of interest to us in this chapter will be first-order reactions. Note that first-order half-life is the only one of the three equations that does not contain a value for $[A]_0$. Therefore, the half-life of a first-order reaction is not dependent on the amount of initial reactant.

3. The half-life for a first-order reaction C → D is known to be 5.0 minutes. After 15.0 minutes have passed, what percentage of C remains? *Hint: How many half-lives have passed?*

4. The rate constant for the first-order reaction A → B is 0.15 s^{-1}. What is the half-life of this reaction? *Hint: Use the half-life equation!*

Review Questions

5. After 2.0 hours, only 25% of a reactant remains. What is the half-life for this reaction?

6. What is the difference in an integrated versus a differential rate law?

7. Write out each integrated rate law, then label *each* variable in *every* equation as either *y*, *m*, *x*, or *b* based on how they relate to a straight line when graphed.

Application Questions

8. The half-life for a zero-order chemical compound is 6.0 hours. How much of a 50.0 g sample will remain after 24.0 hours have passed?

9. The half-life for a first-order chemical compound is 6.0 hours. How much of a 50.0 g sample will remain after 24.0 hours have passed?

10. The half-life for a second-order chemical compound is 6.0 hours. How much of a 50.0 g sample will remain after 24.0 hours have passed?

11. How long, in minutes, will it take for a 30.0 kg sample to decay, through a first-order process, down to 5.0 kg? The half-life for this reaction is 2.0 hours.

12. The half-life of Xanax is 6.0 hours. Let us assume that Xanax decays through a zero-order process. How much (in milligrams) of a 250.0 mg pill will remain after 24.0 hours have passed? *Hint: You will need to use the half-life equation to find the rate constant, then use the integrated rate law to find [A]. You can also use mass in the place of [A]!*

CHAPTER 24

Factors Affecting Rate

What is the process by which reactants convert into products?

What do you call the halfway state between reactant and product?

Learning Outcomes

- Describe intermediates, activated complexes, and transition states.
- Use the collision model of kinetics to describe how temperature and concentration affect the rate of a chemical reaction.
- Describe how a catalyst works.
- Describe a mechanism for a chemical reaction.

Prerequisite Knowledge

- Chapter 22: Kinetics Concepts
- Chapter 23: Integrated Rate Laws

Background Information and Questions

In the previous two chapters, we discussed what rate is and how to determine it as a function of either concentration or time. But why do reactions that look similar on the surface have such different rates? It depends on a number of factors, such as temperature, concentration, state of reactants, the nature of the reactants, and whether a catalyst is present.

The molecules in a reaction must collide with each other for a reaction to occur, but not every collision between molecules will result in a reaction. In order for a reaction to take place, the molecules must not only collide but collide in the correct orientation. **Figure 24.1** illustrates this for a theoretical reaction:

$$2\,AB \rightarrow A_2 + B_2$$

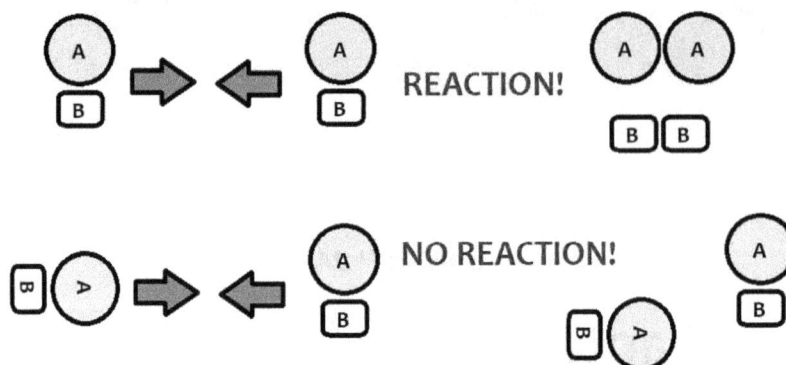

FIGURE 24.1 Requirement of collision for reaction.

These molecules must also collide with kinetic energy greater than the *activation energy* of the reaction. The activation energy (E_a) is the minimum amount of energy necessary to form a product. These values can be related to the rate constant of the reaction using the Arrhenius equation:

$$k = Ae^{-E_a/RT}$$

In this equation, k is the rate constant, A is a term taking into account both the frequency and orientation of collisions, and the $e^{-E_a/RT}$ term accounts the portion of collisions having enough energy to react.

1. Which two things must happen for reactants to react?

2. Would increasing the number of collisions increase or decrease the rate of reaction?

A reaction coordinate diagram shows the progress of a reaction in relation to the energy involved in the reaction. An example is shown in **Figure 24.2**. Note that the x-axis is not time but *progress*, showing the reaction as it goes from beginning to end.

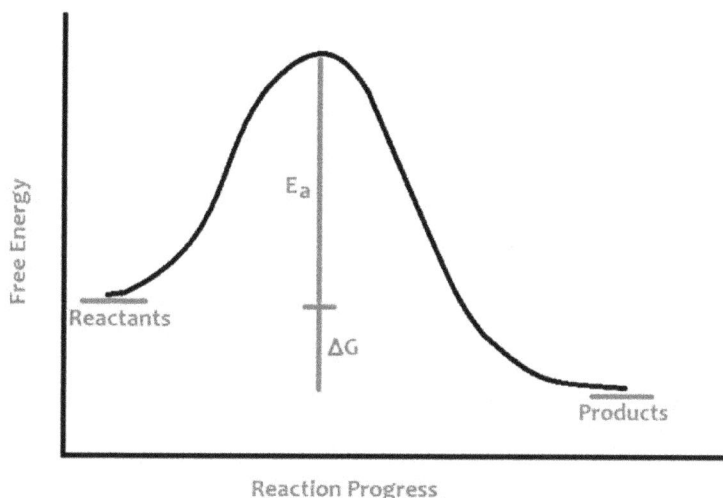

FIGURE 24.2 Reaction coordinate diagram.

The points of interest in this graph are the activation energy and the Gibbs free energy. The activation energy relates to kinetics and how quickly the reaction occurs. Gibbs free energy relates to thermodynamics and the stability and energy of the final product, regardless of how quickly it is formed.

A reaction in which the energy of the products is lower than that of the reactants is said to be *exergonic*, or energy releasing. This term is used interchangeably with exothermic, though they are technically different as exothermic means heat releasing. A reaction in which the energy of the products is higher than that of the reactants is *endergonic*, or energy absorbing. This term is used interchangeably with endothermic, which means heat absorbing, though again these terms do not have the same meaning.

The highest point on this diagram represents the intermediate in the reaction; if there is more than one peak, then there is more than one intermediate. *Intermediates* are formed during the conversion between reactants and products. They are high energy, short lived, and do not appear in the balanced chemical equation. Transition states and activated complexes are two types of intermediates, but we will not go into detail regarding the differences.

One of the hurdles to having a chemical reaction occur is a lack of energy, particularly if the activation energy for the reaction is high. A *catalyst* is a compound that speeds up a reaction by lowering the activation energy. With a lower energy requirement, often because of an alternate pathway, the reaction may proceed more quickly (**Figure 24.3**). Catalysts are not consumed during the reaction, do not affect equilibrium, and do not appear in the balanced chemical equation.

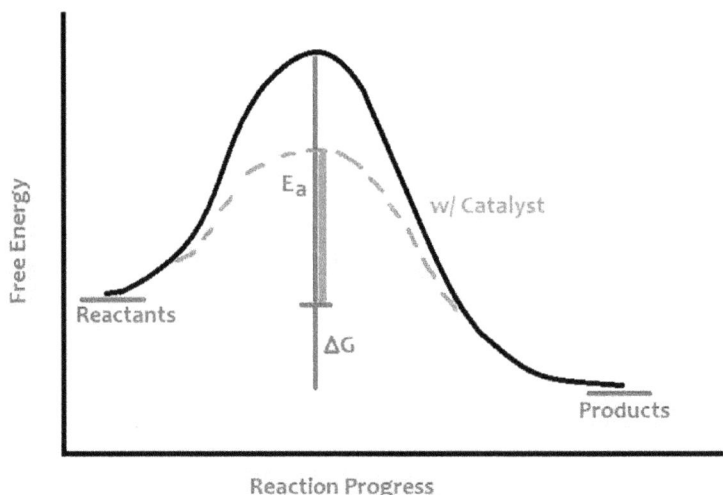

FIGURE 24.3 Reaction coordinate diagram with a catalyst.

3. Can a catalyst be "reused" by the reaction after it speeds up the conversion of one reactant to one product?

If an intermediate doesn't appear in the balanced chemical equation, where does it appear? Well, chemical reactions typically occur in steps. These steps, referred to as the *reaction mechanism*, showcase the process by which a reaction occurs. Each step in the reaction mechanism is called an *elementary reaction*. Consider the following:

Overall reaction: $CO + NO_2 \rightarrow CO_2 + NO$

This reaction actually occurs as two steps, one of which is slow and one of which is fast. Those steps constitute the reaction mechanism and are as follows:

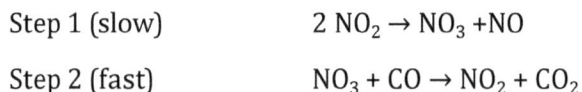

Step 1 (slow) $2\ NO_2 \rightarrow NO_3 + NO$

Step 2 (fast) $NO_3 + CO \rightarrow NO_2 + CO_2$

The intermediate in this case is the NO_3. Note that it is formed in step 1, is used up in step 2, and does not appear in the overall reaction. Because this step is slow, it will determine the overall rate of reaction as well because a reaction can only go as fast as its slowest step. Therefore, we should expect that the overall rate law will include NO_2 (the reactant in the slow step) but perhaps not CO (the reactant in the fast step) because a reaction can only be as fast as its slowest step.

4. Consider the reaction and corresponding elementary steps:

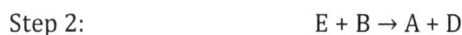

Overall reaction:	$A + B \rightarrow C + D$
Step 1:	$2\,A \rightarrow E + C$
Step 2:	$E + B \rightarrow A + D$

(a) Assume that step 1 is slow and step 2 is fast. Which step is the rate-determining step?

(b) Which of the compounds are intermediates? How do you know?

(c) Which of the reactants do you expect to appear in the rate law, A or B? Why?

Review Questions

5. Do you think the activation energy would be the same for the reverse reaction? Why or why not?

6. How does temperature increase the rate of reaction? *Hint: Which part of the Arrhenius equation does temperature appear in?*

7. How does concentration increase the rate of reaction? *Hint: Having more molecules moving in the same space means that there will be more what? Check the Arrhenius equation!*

8. How and why do you think that factors such as solvent, phase, and surface area might affect the rate of a reaction?

9. If a catalyst acts by providing an alternate path and lowering the activation energy, do you think that any intermediates will be the same or different using a catalyst?

Application Questions

10. Calculate the activation energy for a compound if the frequency factor, A, is found to be $9.50 \times 10^9 \, s^{-1}$ and the rate constant is given as $4.00 \times 10^3 \, s^{-1}$. *Hint: Use the Arrhenius equation!*

11. Calculate the rate constant for a compound if the frequency factor, A, is found to be $9.50 \times 10^9\,s^{-1}$ and the activation energy is given as 39.0 kJ/mol.

12. Given the following reaction and elementary steps, identify the intermediate(s).

    Overall reaction: $2\,NO_2 + Cl_2 \rightarrow 2\,NO_2Cl$

    Step 1: $NO_2 + Cl_2 \rightarrow NO_2Cl + Cl$

    Step 2: $NO_2 + Cl \rightarrow NO_2Cl$

13. If the first step in the previous question is slow, which reactant do you expect to have the most effect on reaction rate?

CHAPTER 25

Equilibrium Concepts

Do reactions always go from reactants toward products?

How does changing one component in a reaction affect the overall reaction?

Learning Outcomes

- Write the equilibrium expression for a chemical reaction.
- Calculate equilibrium constants.
- Predict the direction a reaction will proceed.
- Use Le Châtelier's principle to predict the results from changing conditions in a chemical reaction.

Prerequisite Knowledge

- Chapter 12: Chemical Reactions I
- Chapter 13: Chemical Reactions II
- Chapter 20: Solutions

Background Information and Questions

A common misconception is that chemical reactions involve only the conversion of reactants into products (the forward reaction), while in reality products are also converted into reactants (the reverse reaction). When the rates of these two reactions are equal, the reaction is said to be at *equilibrium*.

Once at equilibrium, the concentration of reactants and products does not change. Reactants are being converted to products at the same rate that products are converted back into reactants. This is represented by a double arrow, as shown in the following reaction:

$$2\,A + B \rightleftharpoons 4\,C$$

From this point forward in the text, we will use a double-headed arrow on most reactions because each reaction has an equilibrium.

    1. What is the forward reaction in the above equation?

    2. What is the reverse reaction in the previous equation?

The *equilibrium constant*, K, describes the position of the equilibrium in a reaction (i.e., whether there will be mostly products or mostly reactants when it reaches equilibrium). K is calculated using the following equation, often summarized as "products over reactants," in which all concentrations are given at equilibrium:

$$aA + bB \rightleftharpoons cC + dD$$

$$K_c = \frac{[C]^c[D]^d}{[A]^a[B]^b}$$

When K > 1 the numerator in the equation must be larger than the denominator, which indicates that there are more products than reactants at equilibrium (product-favored). When K < 1 the denominator must be larger, and therefore, there will be more reactants (reactant-favored).

The value of K is constant for a specific reaction at a specific temperature. Even if different concentrations are involved, the ratio will remain the same.

The subscript following the K denotes the type of equilibrium that is being studied. The example uses K_c in which the "c" stands for *concentration* and molarities are used in the calculation. You may also see K_{eq}, which is just another subscript for K. For an equilibrium expression involving only gases, K_p is often used wherein the "p" stands for "pressure" and the partial pressures of the gases are used instead of concentrations. Additional varieties of K will be discussed in future lessons.

    3. What would be the ratio of products to reactants if K = 1?

4. Calculate K for the following reaction. Assume concentrations at equilibrium are [A] = 0.50 M, [B] = 0.33 M, and [C] = 0.80 M.

$$2A + B \rightleftharpoons 4C$$

Is the forward reaction or the reverse reaction dominant? *Hint: Did you calculate K > 1 or K < 1?*

5. Assume that you have [A] = 2.0 M and [B] = 5.0 M at equilibrium. Use the reaction from the previous questions, along with your calculated value for K, and determine the concentration of [C] at equilibrium.

If the value of K is known for a given reaction, the *direction* of the reaction can be predicted by calculating the *reaction quotient*, Q. The reaction quotient is calculated in the same way as K, but it does not require that the values for concentration be at equilibrium.

Any reaction will always move toward equilibrium because this is most energetically favorable. Therefore, the concentrations of reactants and products will change such that Q moves toward K. For example, if Q = 0.90 and K = 2.2, Q < K and some reactants (in the denominator) must convert to products (in the numerator) to increase the value for Q until it matches K (the reaction will "shift right"). When Q = K the reaction has reached equilibrium. If Q > K, then some products must convert to reactants ("shift left") to lower the value of Q until it reaches that of K.

6. Calculate Q for the following chemical reaction. If K = 0.44, will this reaction shift left or right to reach equilibrium? Concentrations are [A] = 3.0, [B] = 2.0, [C] = 4.0, and [D] = 1.0 for the reaction NOT at equilibrium.

$$A + 3B \rightleftharpoons C + 2D$$

A catalyst will increase the rate of a reaction such that it reaches equilibrium faster, but it does not have an effect on the equilibrium position.

Le Châtelier's principle states that if a system at equilibrium is disturbed or "stressed" by changing conditions, the reaction will return to equilibrium by counteracting the change.

For example, if a product of a chemical reaction at equilibrium was physically removed, that would cause Q to be less than K (because a product, in the numerator, decreased and shifted the reaction away from equilibrium). Because K is a constant, it cannot change, so instead, the reaction will shift to produce more products to make up for those that were lost (a shift right, toward the products).

This is a useful industrial application of equilibrium because the reaction can be manipulated to favor products. Consider the following example:

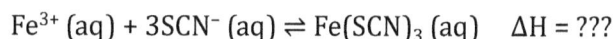

$$Fe^{3+} (aq) + 3SCN^- (aq) \rightleftharpoons Fe(SCN)_3 (aq) \quad \Delta H = ???$$

The iron(III) ion is yellow in color, the thiocyanate ion is clear, and the iron(III) thiocyanate is red (**Figure 25.1**). The equilibrium constant for the reaction is given as:

$$K_c = \frac{[Fe(SCN)_3]}{[Fe^{3+}][SCN^-]^3}$$

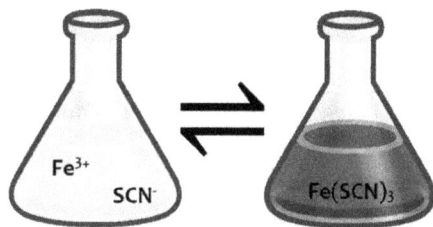

FIGURE 25.1 Color shift between yellow (Fe^{3+} and SCN^-) and red ($Fe(SCN)_3$) with reaction.

If some of the iron(III) was removed via precipitation, then a reactant will have been removed, Q will be larger than K, and some products will convert to reactants to regain equilibrium. This would be a shift to the left.

When determining how temperature affects equilibrium, it is important to determine whether a reaction is endothermic or exothermic. If the temperature of this reaction were increased, the color would shift more toward yellow (toward reactants or toward the left), indicating that heat is a product and this reaction is exothermic.

A summary of Le Châtelier's principle is provided in **Table 25.1**.

TABLE 25.1 Le Châtelier's Principle

Change	Shift	Why
Reactant is added.	Right	Q→K
Reactant is removed.	Left	Q→K
Product is added.	Left	Q→K
Product is removed.	Right	Q→K
Pressure is increased (gases).	Toward side with fewer gas molecules.	An increase in pressure leads to a decrease in volume.
Pressure is decreased (gases).	Toward side with more gas molecules.	A decrease in pressure leads to an increase in volume.

Whether a reaction is endothermic (absorbs heat) or exothermic (evolves heat) determines the effect of changing the temperature. For an endothermic reaction heat is a reactant, and for an exothermic reaction heat is a product.

7. Consider the reaction used in the production of ammonia:

$$N_2 \text{ (g)} + 3 H_2 \text{ (g)} \rightleftharpoons 2 NH_3 \text{ (g)}$$

(a) If the ammonia is removed, which way will the reaction shift, left or right?

(b) If the pressure is decreased, which way will the reaction shift, left or right?

Review Questions

Refer to the following reaction and diagrams to answer the questions.

@ Equilibrium

 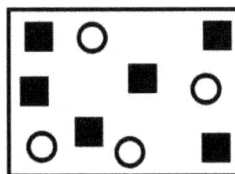

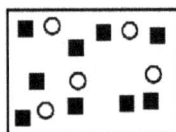 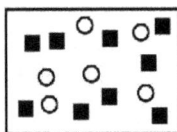 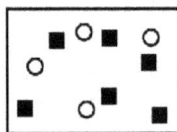 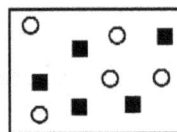

I   II   III   IV

8. If the pressure of the system at equilibrium is increased, which box will best represent the new equilibrium? Why?

9. If some ■ is added to the system when it is at equilibrium, which box will represent the new equilibrium? Why?

10. If the temperature of this reaction is decreased, which box will represent the new equilibrium? Why?

11. Explain why equilibrium is dynamic instead of static. Does a reaction ever "stop," even when concentrations are stable?

Application Questions

12. The Haber process is an industrial process for producing ammonia from hydrogen and nitrogen gases. It is a vital process that is used in production of polymers, fertilizer, and various other products. In industry this reaction is done at high temperature and pressure with a catalyst. The basic reaction is:

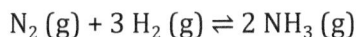

$$N_2 \, (g) + 3 \, H_2 \, (g) \rightleftharpoons 2 \, NH_3 \, (g)$$

Given the following information at equilibrium, determine the partial pressure of ammonia at equilibrium: $K_p = 1.640 \times 10^{-4}$ at 400 °C, $P_{H_2} = 100.0$ atm, $P_{N_2} = 150.0$ atm. *Hint: Recall that you can use partial pressures instead of concentrations to calculate K!*

13. The Haber process is often done at pressures of 400 atm or higher in indus-try. Assuming that the goal is to produce ammonia efficiently, why is higher pressure better?

14. The forward reaction in the Haber process is exothermic. Will a high or a low temperature favor ammonia production?

15. The Haber process is typically conducted at temperatures of around 400 °C to 450 °C. This is a fairly high temperature. Explain this in the context of your answer to the previous question.

16. Phosphorus pentachloride decomposes in the presence of water to produce heat, which may ignite surrounding materials. Production of PCl_5 is given by the following:

$$PCl_3 + Cl_2 \rightleftharpoons PCl_5$$

The equilibrium constant, K_c, at a certain temperature is known to be 0.030. If you have the following concentrations of each compound, will this reaction proceed to the left or to the right to reach equilibrium?

[PCl_3] = 4.8 M   [Cl_2] = 3.4 M   [PCl_5] = 0.90 M

CHAPTER 26

Acids and Bases

What do the terms "acidic" and "alkaline" mean?

Are all acids dangerous?

Learning Outcomes

- Identify acids, bases, and conjugate pairs.
- Calculate pH and pOH.
- Predict products of acid–base reactions.

Prerequisite Knowledge

- Chapter 12: Chemical Reactions I
- Chapter 13: Chemical Reactions II
- Chapter 25: Equilibrium Concepts

Background Information and Questions

Acids and bases have several definitions that can be applied depending on the type of chemistry being studied. Several such definitions are listed in **Table 26.1**.

The definition we will focus on in this chapter is the Brønsted-Lowry definition. After an acid loses (or donates) its proton, what is left is the conjugate base of that acid. The base accepts the proton, and the newly formed compound is the conjugate acid of that base.

TABLE 26.1 Definitions of Acids and Bases

Term	Definition
Arrhenius acid	Produces H⁺ ions in aqueous solution
Arrhenius base	Produces OH⁻ ions in aqueous solution
Brønsted–Lowry acid	Donates a proton (H⁺ ion) in solution
Brønsted–Lowry base	Accepts a proton (H⁺ ion) in solution
Lewis acid	Electron acceptor
Lewis base	Electron donor

In the following example, CH_3COOH is the acid because it loses a proton during the reaction. Sodium hydroxide is the base because it gains the proton that is lost. The conjugate base is CH_3COO^-, which is in the form of a salt with Na^+, and water is the conjugate acid. Some compounds, such as water, can act as either an acid or a base and are referred to as being *amphoteric*.

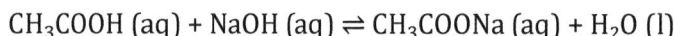

$$CH_3COOH \text{ (aq)} + NaOH \text{ (aq)} \rightleftharpoons CH_3COONa \text{ (aq)} + H_2O \text{ (l)}$$

1. Label the acid, base, conjugate acid, and conjugate base for the following:

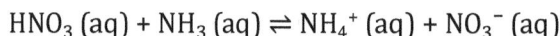

$$HNO_3 \text{ (aq)} + NH_3 \text{ (aq)} \rightleftharpoons NH_4^+ \text{ (aq)} + NO_3^- \text{ (aq)}$$

Various equilibria are associated with acids and bases. One common use of equilibrium constants is to determine the strength of an acid using the acid dissociation constant, K_a, which is simply the equilibrium constant for the dissociation of an acid. K_b is the analogous constant for a base. Acids are often represented by a generic HA in the place of an actual acid, with the understanding that HA is the acid, H^+ is the proton, and A^- is the conjugate base. The acid dissociation constant, K_a, is therefore:

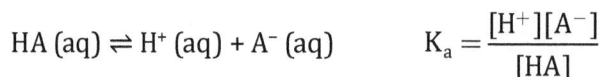

$$HA \text{ (aq)} \rightleftharpoons H^+ \text{ (aq)} + A^- \text{ (aq)} \qquad K_a = \frac{[H^+][A^-]}{[HA]}$$

2. Write the acid dissociation constant, K_a, for acetic acid in the following equation. *Note: Pure liquids and solids are not included in any equilibrium expressions, so H_2O (l) does not go into the equation, only compounds or ions in aqueous solution (aq).*

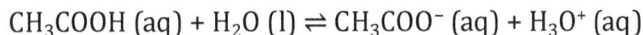

$$CH_3COOH \text{ (aq)} + H_2O \text{ (l)} \rightleftharpoons CH_3COO^- \text{ (aq)} + H_3O^+ \text{ (aq)}$$

In the previous question, H_3O^+ (hydronium ion) was produced as the conjugate acid. Although the hydronium ion is not the same as the hydrogen ion, they are often used interchangeably in acid–base equilibrium expressions because both ions represent the number of protons released into aqueous solution. The generic reaction for an acid dissociation including water is illustrated below.

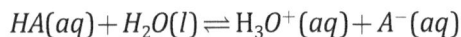

$$HA(aq) + H_2O(l) \rightleftharpoons H_3O^+(aq) + A^-(aq)$$

Because acids are defined as compounds that donate protons in solution, it follows that the more protons an acid releases into solution the stronger an acid it is. The acid dissociation constant, K_a, can therefore be used to measure acid strength. A larger K_a means a stronger acid because the numerator, which accounts for the donated protons, will be larger compared to the undissociated acid in the denominator.

3. Rank the following from *weakest* to *strongest* acid:

Boric acid, H_3BO_3 $K_a = 5.81 \times 10^{-10}$

Iodic acid, HIO_3 $K_a = 1.7 \times 10^{-1}$

Nitrous acid, HNO_2 $K_a = 7.1 \times 10^{-4}$

Acids that have measurable equilibria in water are said to be "weak" acids, even if some are stronger than others. A "strong" acid is one that dissociates completely in water, making it impossible to measure K_a in aqueous solution. Similarly, some bases will dissociate completely, with the reaction essentially going to completion rather than reaching an equilibrium. Hydrochloric acid is one such example:

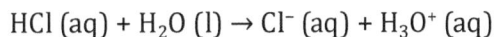

$$HCl \text{ (aq)} + H_2O \text{ (l)} \rightarrow Cl^- \text{ (aq)} + H_3O^+ \text{ (aq)}$$

Notice the lack of a double-headed equilibrium arrow. This reaction goes to completion, producing 100% products. Hydrochloric acid is therefore considered to be "strong." Other common strong acids and bases are listed in **Table 26.2**.

TABLE 26.2 Common "Strong" Acids and Bases

Acids	Bases
HCl	LiOH
HBr	NaOH
HI	KOH
HNO_3	RbOH
$HClO_3$	CsOH
$HClO_4$	$Ca(OH)_2$
H_2SO_4	$Sr(OH)_2$
	$Ba(OH)_2$

Water, as an amphoteric species, is capable of undergoing *autoionization* by reacting with itself in a reaction where one molecule of water is an acid and one is a base. In pure water at 25 °C, the equilibrium constant for this reaction is given as K_w:

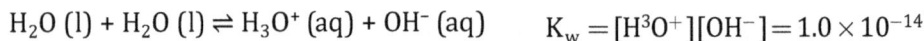

$$H_2O \ (l) + H_2O \ (l) \rightleftharpoons H_3O^+ \ (aq) + OH^- \ (aq) \qquad K_w = [H^3O^+][OH^-] = 1.0 \times 10^{-14}$$

Rather than judge the strength of the acid or base itself, the concentrations of hydronium and hydroxide ions in an aqueous solution determine whether the solution is acidic or basic (alkaline). A solution is acidic if it contains more hydronium than hydroxide, basic if it contains more hydroxide than hydronium, and neutral if it contains equal amounts of each.

Listing these amounts in molarities is cumbersome due to the use of scientific notation and the large range of values possible; therefore, a logarithmic scale known as the *pH scale* is used. The "p" indicates a negative logarithm, and it can be shown that the pH scale goes from 0 to 14 because $pK_w = -\log(K_w) = 14$. Other relevant equations and information are shown in **Table 26.3**.

TABLE 26.3 pH Scale Equations

pH Scale
pH = −log(H⁺) = −log(H₃O⁺)
pOH = −log(OH⁻)
pH + pOH = 14
Acidic solution: pH < 7
Basic solution: pH > 7
Neutral solution: pH = 7

Here's an example using a 0.35 M solution of HBr, a strong acid. Because HBr is a strong acid, it will fully dissociate in solution, giving 100% products in the following reaction:

$$HBr + H_2O \rightarrow H_3O^+ + Br^-$$

A 0.35 M solution of HBr therefore produces 0.35 M H_3O^+, which can be used to calculate the pH:

$$pH = -\log[H_3O^+] = -\log(0.35) = 0.46$$

The solution is clearly an acid, having a pH significantly lower than 7. Once the pH is obtained, it is easy to determine the pOH; just subtract 0.46 from 14 to get pOH = 13.54. If you needed the [OH⁻] concentration, you could calculate that as well as follows:

$$pOH = -\log[OH^-]$$

$$13.54 = -\log[OH^-]$$

$$[OH^-] = 2.89 \times 10^{-14} \text{ M}$$

4. Calculate the pH of a solution that has [H⁺] = 2.0×10^{-4} M.

   (a) What is the pOH of the solution?

   (b) Is this solution acidic or basic (alkaline)?

5. Calculate the [H⁺] concentration of a solution that has a pH = 8.5.

6. Calculate the pH of a 0.3 M HNO_3 solution. *Hint: Recall that HNO_3 is a strong acid and will fully dissociate! If all of the reactants are converted into products, how much H^+ will be present?*

Calculating the pH of a weak acid or base is more complicated than for a strong acid because not all of the acid (or base) is converted into H_3O^+ or OH^-. In this case the acid dissociation constant, K_a, is used alongside something called an ICE table.

ICE is an acronym for "initial, change, equilibrium," and it is used to determine the equilibrium concentrations of the reactants and products in solution. Once the equilibrium concentration of H_3O^+ is known, the pH can be calculated. Consider the following example.

What is the pH of a 2.60 M solution of formic acid, HCOOH? $K_a = 1.80 \times 10^{-4}$

To begin, write out the reaction that is occurring, and then draw a blank ICE table as shown in **Table 26.4** and list the reactants and products above each column. The concentration of water will not change significantly and may be left blank; recall that liquids and solids are not part of the equilibrium expression.

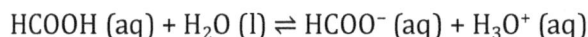

$$HCOOH \ (aq) + H_2O \ (l) \rightleftharpoons HCOO^- \ (aq) + H_3O^+ \ (aq)$$

TABLE 26.4 ICE Table: Step 1

		[HCOOH]	**[H₂O]**	**[HCOO⁻]**	**[H₃O⁺]**
Initial	**I**		—		
Change	**C**		—		
Equilibrium	**E**		—		

Next, fill in all the information you have from the problem. In this example we only know the initial concentrations are 2.60 M for formic acid and 0 for both of the products, since the reaction has not started yet (**Table 26.5**).

TABLE 26.5 ICE Table: Step 2

		[HCOOH]	**[H₂O]**	**[HCOO⁻]**	**[H₃O⁺]**
Initial	**I**	2.60	—	0	0

			—		
Change	**C**		—		
Equilibrium	**E**		—		

Now check that your chemical equation is balanced and determine how many of the acid molecules will react to produce how many products. In this example, one molecule of HCOOH (formic acid) will react with water to produce one molecule of HCOO⁻ and one molecule of H_3O^+. So every "x" amount of formic acid that reacts will produce the same amount of each product; that is, the "change" that is indicated by a loss of reactants and a gain of products (**Table 26.6**). This will not always be a 1:1 ratio; the reaction will depend on the stoichiometric coefficients in the reaction.

TABLE 26.6 ICE Table: Step 3

		[HCOOH]	[H₂O]	[HCOO⁻]	[H₃O⁺]
Initial	**I**	2.6	—	0	0
Change	**C**	–x	—	+x	+x
Equilibrium	**E**		—		

The equilibrium concentrations can now be calculated, as shown in **Table 26.7**.

TABLE 26.7 ICE Table: Step 4

		[HCOOH]	[H₂O]	[HCOO⁻]	[H₃O⁺]
Initial	**I**	2.6	—	0	0
Change	**C**	–x	—	+x	+x
Equilibrium	**E**	2.6 – x	x	x	x

Now that equilibrium concentrations are known for each compound, the acid dissociation constant can be used to solve for x.

$$K_a = \frac{[HCOO^-][H^3O^+]}{[HCOOH]} = \frac{[x][x]}{[2.6-x]} = \frac{x^2}{2.6-x}$$

$$K_a = 1.8\times10^{-4} = \frac{x^2}{2.6-x}$$

At this point the quadratic equation can be used to solve for x, which will be equal to the concentration of H_3O^+ (see the ICE table in Table 26.7 for why). However, if we assume that the value for x is much, much less than 2.6, we can estimate that $2.6 - x$ is approximately equal to 2.6. The problem then becomes simple to solve:

$$K_a = 1.8 \times 10^{-4} = \frac{x^2}{2.6 - x}$$

$$x = 0.022 = [H^3O^+]$$

$$pH = -\log(0.022) = 1.66$$

The rule for ignoring the minus x is if the percent ionization is less than 5%, then the $-x$ can be ignored.

7. What is the pH of a solution that is 0.40 M acetic acid, CH_3COOH? $K_a = 1.75 \times 10^{-5}$

Review Questions

8. Rank the following acids in order from *strongest* to *weakest*.

Lactic acid	$pK_a = 3.08$
Acetylsalicylic acid	$pK_a = 3.48$
Oxalic acid	$pK_a = 1.23$
Ascorbic acid	$pK_a = 4.10$

9. Why is pK_a used more often than K_a to judge acid strength?

Application Questions

10. What is the pH of a solution that is 0.044 M nitrous acid, HNO_2? $K_a = 7.1 \times 10^{-4}$ for HNO_2

11. The concept of an ICE table can be applied to calculating values for equilibria that are not associated with acids or bases. Consider the following:

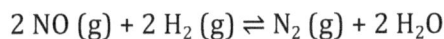

$$2\ NO\ (g) + 2\ H_2\ (g) \rightleftharpoons N_2\ (g) + 2\ H_2O$$

A mixture of 2.0 M NO, 1.0 M H_2, and 2.0 M H_2O reaches equilibrium (there is no N_2 at the start of the reactions). At equilibrium, the concentration of NO is 1.2 M.

Calculate K_c for this reaction,

(a) Initial concentrations are [CO] = 1.5 M, [H_2O] = 3.0 M.

(b) At equilibrium the concentration of [CO_2] = 2.13 M.

CHAPTER 27

Applying Equilibria

What is a "salt," exactly?

We eat foods containing acids and bases, so how do our bodies stay the same pH?

How can we find the concentration of an unknown acid sample?

Learning Outcomes

▪ Predict whether a salt solution will be acidic, basic, or neutral.
▪ Calculate the concentration of an unknown using titration.
▪ Identify buffers and buffer regions on a titration curve.
▪ Prepare a buffer of a given pH.

Prerequisite Knowledge

▪ Chapter 26: Acids and Bases

Background Information and Questions

Recall that a neutralization reaction can be defined as *acid plus base equals salt plus water*. But what is a salt, exactly? A salt can be defined as an ionic compound that results from the neutralization reaction of an acid and a base. For example:

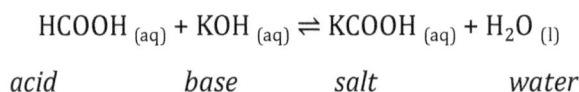

$$\text{HCOOH}_{(aq)} + \text{KOH}_{(aq)} \rightleftharpoons \text{KCOOH}_{(aq)} + \text{H}_2\text{O}_{(l)}$$

acid   *base*   *salt*   *water*

Because a salt is an ionic compound, it will exist in solution as ions. In the example, the two ions would be HCOO^- and K^+. We can think of these ions as conjugates of either acids or bases. The conjugate base of a strong acid will be a very weak base, whereas the conjugate base of a weak acid will be a stronger base (though not "strong").

H_2O is the conjugate acid of KOH, a strong base; therefore, K^+ will be a very weak acid and have a negligible effect on the pH. $HCOO^-$ is the conjugate base of a weak acid, however, and will therefore be a stronger base than K^+ is an acid. The result is that if a sample of the salt HCOOK (s) was placed into aqueous solution, the pH would be greater than 7, indicating a basic solution.

1. Determine whether each of the following salts would produce an acidic, basic, or neutral solution.

$NaCl$ $\qquad\qquad$ NH_4Cl $\qquad\qquad$ $NaCN$

2. A reaction between a strong base and a weak acid will produce a weakly basic solution. How is this related to the strength of the conjugate acid and base?

Neutralization reactions can also be used to determine the concentration of an unknown acid or base in a *titration*. In a titration, a known concentration of *titrant* is used to determine the concentration of the *analyte* using stoichiometry. The exact amount of titrant needed to react with the analyte is the *equivalence point* of the titration. Dyes called *indicators*, which are often weak acids that are one color in acidic media and another color in basic, are used to obtain an estimate of the equivalence point, called the *end point*. For example:

A 30.0 mL sample of NaOH is titrated with a solution that is 2.0 M H_2SO_4. The end point is reached after 15.8 mL of H_2SO_4 are added. What is the concentration of the NaOH? Begin by writing the relevant chemical equation and balancing it.

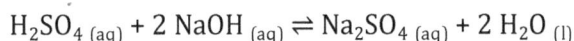

$$H_2SO_{4\ (aq)} + 2\ NaOH_{\ (aq)} \rightleftharpoons Na_2SO_{4\ (aq)} + 2\ H_2O_{\ (l)}$$

When the end point is reached, it means that enough sulfuric acid has been added to react with every molecule of NaOH. Assuming that the titration is stopped at the end point, no excess acid is present. Therefore, the amount of acid added is stoichiometrically equivalent to the amount of base present, and we can solve for the unknown concentration.

$$\frac{2.0\ \text{mol}\ H_2SO_4}{1\ L} \times \frac{0.0158\ L}{1} \times \frac{2\ \text{mol NaOH}}{1\ \text{mol}\ H_2SO_4} \times \frac{1}{0.03\ L} = 2.10\ \text{M NaOH}$$

3. A 20.0 mL sample of HCl with unknown concentration is titrated with 22.5 mL of 0.5 M NaOH to reach the end point. What is the concentration of HCl in the unknown solution?

4. Why is it important to stop a titration as close to the equivalence point as possible?

The sulfuric acid mentioned in the previous example is a *polyprotic acid*. Polyprotic acids are those that have more than one hydrogen (or proton) to lose. Sulfuric acid has two hydrogens, phosphoric acid has three, hydrochloric has one, and so one. Many weak acids, such as acetic acid (CH_3COOH), may appear to have more than one hydrogen to lose, but the hydrogens attached to carbon atoms are not acidic hydrogens and are not lost during acid–base reactions.

Because polyprotic acids have more than one hydrogen to lose, they have more than one value for the acid dissociation constant, K_a. **Figure 27.1** shows what happens at each point on a titration curve for a triprotic acid.

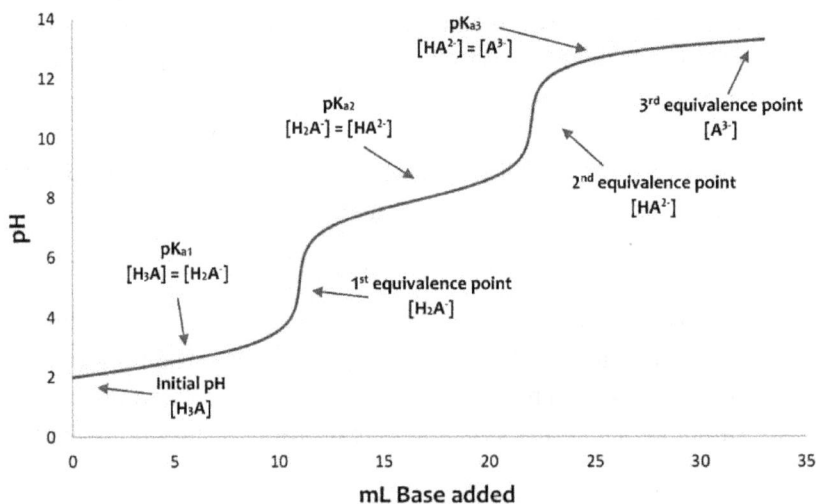

FIGURE 27.1 Polyprotic acid titration curve.

Note that pK_a (the negative log of K_a) is used instead because the value for pK_a correlates to pH. Because base is added and the pH increases, hydrogen ions are removed one by one from the acid. Each subsequent hydrogen is less acidic than the first as they become harder to remove.

5. Would a smaller pK_a indicate a stronger or a weaker acid? Hint: Recall that a larger K_a indicated a stronger acid.

Note that on the titration curve, the pK_a values represent "flat" areas of the curve where the pH changes slowly and the amount of acid present is equal to the amount of its conjugate base. These areas, within ±1 pH of the pK_a, are called *buffer regions*.

A *buffer* is a mixture of a weak acid and its conjugate base (or a weak base and its conjugate acid). A buffer will resist changes in pH because any acid added will react with the base present in solution, and any base added will react with the acid present in solution. A buffer with higher concentrations of weak acid and conjugate base is said to have a higher *buffer capacity*, meaning that more acid or base can be added before a change in pH is measured.

In order to add a conjugate base to solution, a salt is typically used. In the previous example, the buffer could be prepared with a mixture of the weak acid CH_3COOH (acetic acid) and the salt CH_3COONa (sodium acetate). In solution the CH_3COONa would separate into Na^+ (the conjugate acid of a strong base, which has little effect on pH) and CH_3COO^- (the conjugate base of the weak acid used in the buffer).

For example, consider what will happen when a strong acid or base is added to an acetic acid buffer consisting of CH_3COOH (acetic acid) and its conjugate base CH_3COO^- (acetate ion):

$$\text{Add NaOH:} \quad NaOH + CH_3COOH \rightleftharpoons CH_3COO^-Na^+ + H_2O$$

In this case the conjugate base (already present in solution) is generated along with water, maintaining the pH of the buffer.

$$\text{Add HCl:} \quad HCl + CH_3COO^-Na^+ \rightleftharpoons CH_3COOH + Na^+Cl^-$$

In this case the weak acid (already present in solution) is generated along with sodium chloride ions, which have little effect on pH, again maintaining the pH of the buffer.

A buffer will work best when the amounts of weak acid and conjugate base are equal, which occurs when the pH of the buffer equals the pK_a of the acid. For an acetic acid buffer:

$$K_a(\text{acetic acid}) = 1.75 \cdot 10^{-5}$$

$$pK_a = -\log(K_a) = -\log\left(1.75 \cdot 10^{-5}\right) = 4.76$$

Therefore, an acetic acid buffer would work best at pH = 4.76. By adjusting the amounts of the weak acid and conjugate base in the buffer, it is possible to prepare a buffer that will work between 3.76 and 5.76 ($\pm$1 pH of the pK_a).

6. Which of the following would be the best choice to prepare a buffer at pH = 9?

HNO_2 $K_a = 4.0 \times 10^{-4}$

HCN $K_a = 6.17 \times 10^{-10}$

H_2O_2 $K_a = 2.4 \times 10^{-12}$

H_2S $K_a = 9.1 \times 10^{-8}$

7. Using the data provided in the previous question, what pH range could HNO_2 be an effective buffer for?

If adding a weak acid and its conjugate base in a 1:1 ratio provides a buffer at pH = pKa, how can a buffer at a different pH be prepared? In this case the Henderson–Hasselbalch equation is used.

$$pH = pK_a + \log\left(\frac{[A^-]}{[HA]}\right)$$

Recall that HA is used to denote a generic weak acid and A^- its conjugate base. The equation can be used to determine the pH of a specific buffer. Consider the following example. What is the pH of a buffer solution made with 0.20 M HF and 0.35 M NaF?

The pK_a of HF is 3.14.

First, determine the identities and concentration of the weak acid (HA) and the conjugate base (A^-). In this case the weak acid is 0.20 M HF and the conjugate base is 0.35 M F^- (because NaF will produce F^- ions in solution in a 1:1 ratio). Then plug the data into the equation:

$$pH = pK_a + \log\left(\frac{[A^-]}{[HA]}\right) = 3.14 + \log\left(\frac{[0.35]}{[0.20]}\right) = 3.38$$

This mixture would produce a buffer of pH = 3.38. This makes sense because if we used a higher concentration of base than acid, the buffer pH would be slightly higher than the pK_a of 3.14.

8. Calculate the pH of a buffer composed of 0.41 M HCN and 0.33 M KCN. Refer to previous questions for the K_a (and pK_a) of HCN.

Review Questions

9. In this chapter, we discussed creating a buffer by mixing a weak acid with its conjugate base in salt form. It is also possible to create a buffer by mixing an acid with a strong base:

$$0.5 \text{ M KOH} + 1.0 \text{ M HA} \rightarrow \text{Buffer}$$

Draw out the chemical reaction taking place, and explain why this would work to generate a buffer solution.

10. How do salts affect the pH of a solution when they are neither acids nor bases?

Application Questions

11. A 25.0 mL sample of KOH is titrated with 40.0 mL of 0.77 M HCl. What is the concentration of the unknown solution?

12. What is the pH of a buffer prepared by combining 0.020 mol acetic acid with 0.030 mol sodium acetate? $K_a = 1.75 \times 10^{-5}$

13. Which of the following buffer systems would be the best choice to create a buffer of pH = 9.0?

    (a) CH_3COOH/CH_3COOK

    (b) $HClO_2/KClO_2$

    (c) NH_3/NH_4Cl

14. Refer to the previous question. Which compound in each pair is a salt? Would adding each of these salts to solution produce an acidic, alkaline, or neutral solution?

CHAPTER 28

Additional Equilibria

What other kinds of equilibria exist?

Why might less solid dissolve in a salt solution compared to pure water?

Learning Outcomes

- Write equilibrium constants for solubility equilibria.
- Calculate solubility.
- Predict effects caused by a common ion in solution.
- Define other types of equilibrium constants.

Prerequisite Knowledge

- Chapter 25: Equilibrium Concepts
- Chapter 26: Acids and Bases
- Chapter 27: Applying Equilibria

Background Information and Questions

In the previous chapter, salts—a combination of an anion and a cation created in a neutralization reaction—were discussed. Salts have a tendency to dissociate in aqueous solution. The equilibrium constant for this process is K_{sp}, known as the *solubility product constant*, and represents the equilibrium between the undissolved solid and the ions in solution. This makes K_{sp} representative of a *heterogeneous* equilibrium, one that involves compounds in two different states (solid and aqueous in this case). A *homogeneous* equilibrium would have all substances in the same state.

Using the dissociation of barium fluoride as an example:

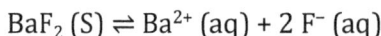

$$BaF_2 \text{ (S)} \rightleftharpoons Ba^{2+} \text{ (aq)} + 2 \text{ F}^- \text{ (aq)}$$

$$K_{sp} = [Ba^{2+}][F^-]^2$$

1. Write the solubility product constant (K_{sp}) expressions for the following. *Recall that solids are not included in equilibrium expressions!*

   (a) $AlPO_4$

   (b) $AgCl$

   (c) Na_2SO_4

   (d) $Ca(NO_3)_2$

The degree to which a particular salt dissociates is known as the *solubility* of the salt, *s*. The relationship between the salt, the K_{sp}, and the ions can be used to calculate the solubility. Consider the following example:

The K_{sp} for barium fluoride is 1.84×10^{-7}. What is the solubility of barium fluoride?

To begin, look at the molar relationship between the salt and each of its ions in the balanced chemical equation. There is a 1:1 molar ratio between BaF_2 and Ba^{2+}; therefore, $Ba^{2+} = S$. There is a 1:2 molar ratio between BaF_2 and F^-; therefore, there are twice as many moles of fluoride ion in solution as there are moles of solid barium fluoride, so $F^- = 2S$. These values can then be plugged into the K_{sp} equation to solve for S:

$$K_{sp} = [Ba^{2+}][F^-]^2$$

$$1.84 \times 10^{-7} = [s][2s]^2$$

$$s = 0.00358 \text{ M} = 3.58 \times 10^{-3}$$

Therefore, a maximum of 0.00358 moles of barium fluoride will dissolve per liter of water at a given temperature and pressure.

2. What is the solubility of $Ca(OH)_2$? $K_{sp} = 5.02 \times 10^{-6}$

3. What is the solubility of $CaCO_3$? $K_{sp} = 3.36 \times 10^{-9}$

4. Which of the two compounds, $CaCO_3$ or $Ca(OH)_2$, is more soluble in water? Explain.

Similarly, if the concentration of ions in solution can be determined, the K_{sp} for the solid can be found by working backward. Consider the following example:

What is the K_{sp} of AgBr given that the concentration of Ag^+ at equilibrium is 5.85×10^{-7} M?

In this case, $Ag^+ = Br^- = 5.85 \times 10^{-7} = S$ (all molar ratios in this example are 1:1).

$$K_{sp} = [Ag^+][Br^-] = [5.85 \times 10^{-7}][5.85 \times 10^{-7}] = 3.42 \times 10^{-13}$$

5. What is the K_{sp} of AgCN if the molar solubility is 7.65×10^{-9} M?

When a solid, salt compound is dissolved into solution it produces ions, but what if one (or both) of those ions are already present in solution? In that case the *common ion effect*, predicted by Le Châtelier's principle, *decreases* the solubility of the solid compound.

For example, the presence of chloride ion in a solution will decrease the amount of AgCl that can dissolve in that solution, thereby lowering the solubility. In this case the Cl^- would be considered the "common ion." Because it is a product of the dissolution of AgCl, it will shift that reaction back toward the reactants (in this case, the solid).

Calculate the solubility of AgCl in water and in a solution that has $[Cl^-] = 0.02$ M if the K_{sp} for AgCl is 1.75×10^{-10}.

$$K_{sp} = 1.75 \times 10^{-10} = [Ag^+][Cl^-]$$

In pure water: $1.75 \times 10^{-10} = [s][s] \rightarrow s = 1.32 \times 10^{-5}$

In 0.02 M Cl^-: $1.75 \times 10^{-10} = [s][s+0.02] \rightarrow s = 8.75 \times 10^{-9}$

The trick! In most cases where the concentration of the common ion is within a factor of 104 of the solubility, the (S + 0.02) part of the equation can be estimated as (0.02). This makes the equation much easier to solve:

$$1.75 \times 10^{-10} = [s][0.02] \rightarrow s = 8.75 \times 10^{-9}$$

The amount of Cl^- that is added from the AgCl is so small compared to the Cl^- already in solution that it is considered negligible!

6. What is the solubility of AgBr in a 0.150 M solution of NaBr? Assume the K_{sp} for AgBr is 3.50×10^{-13}.

Review Questions

7. Explain how the concept of the "common ion" might work in terms of buffer solutions.

8. The solubility of a compound cannot simply be ranked by the K_{sp} value because acids can be ranked by their K_a values since K_{sp} depends on the total number of ions that the salt produces in solution. With this in mind, why is it technically okay to rank $Ca(NO_3)_2$ and $Mg(OH)_2$ by their K_{sp} values but NOT $Ca(NO_3)_2$ and KNO_3?

Application Questions

9. How many moles of AgCl would dissolve in 1 L of water?

$$AgCl \ K_{sp} = 1.75 \times 10^{-10}$$

10. How many moles of AgCl would dissolve in 200.0 mL of water?

11. How many grams of AgCl would dissolve in 200.0 mL of water?

12. What is the K_{sp} of magnesium phosphate if the concentration of PO_4^{3-} in solution at equilibrium is 7.88×10^{-6} M?

$$Mg_3(PO_4)_{2 \ (s)} \rightleftharpoons 3Mg^{2+}_{\ (aq)} + 2PO_4^{3-}_{\ (aq)}$$

CHAPTER 29

Energetics I

What is energy?

Can we create energy?

How do we determine the amount of heat gained or lost in a process?

Learning Outcomes

- Describe and illustrate the difference between potential and kinetic energy.
- Calculate changes in internal energy using heat and work.
- Compute any of the variables in $q = mc\Delta T$.
- Explain the principles of calorimetry and calculate energies released or absorbed in a calorimeter.

Prerequisite Knowledge

- Chapter 12: Chemical Reactions I
- Chapter 13: Chemical Reactions II
- Chapter 14: Introduction to Stoichiometry

Background Information and Questions

Energy plays a significant role in chemistry; it can cause chemical reactions or be produced by them. The formal definition of *energy* is the ability to do work or transfer heat. In the equation for energy below, q = heat and w = work:

$$E = q + w$$

Energy is a difficult concept to visualize because you cannot see energy. However, you can see the effects and conversion of energy. Your automobile is an energy conversion device; it converts chemical energy in the bonds of gasoline to mechanical energy. Chemical reactions, like the one happening in your car's engine, convert energy as well.

The *first law of thermodynamics* states that energy cannot be created or destroyed; it is simply transformed or converted to other forms.

Most of this chapter will focus on energy related to chemical reactions. In this chapter you will see the terms *thermodynamics* and *thermochemistry* quite a bit. Thermodynamics is the study of energy and its transfers and conversions; thermochemistry is a division of thermodynamics that focuses on chemical reactions and energy, more specifically thermal energy.

The two main types of energy are kinetic energy and potential energy; however, it has many forms, including chemical, nuclear, mechanical, electrical, and so on. Kinetic energy is associated with motion, whereas potential energy is often called "stored energy." All forms of energy can be classified as kinetic or potential. **Figure 29.1** illustrates the types of energy and some examples of the energy forms. Energy is converted between these types and forms.

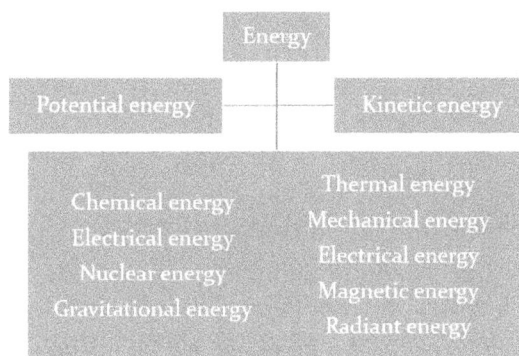

FIGURE 29.1 Types of energy.

Potential energy can be conceptualized by thinking about the force of gravity as a function of height. Think of a diver (**Figure 29.2**). On the diving platform, all of the energy is potential. As the diver begins to dive, the energy is converted from potential to kinetic. A similar conversion occurs with many energy-involved processes.

1. Using a roller coaster as an example, illustrate and describe points on a roller coaster where energy will be all potential, all kinetic, and a mixture of the two.

FIGURE 29.2 Conversion of energy in a Dive.

When studying energy, we typically refer to the object under study as the *system*. Everything outside the system is known as the *surroundings*. For example, a bowl of soup could be your system; the atmosphere around it is the surroundings. It is important to note that E is the total energy (kinetic + potential) in a system and is known as *internal energy* (E). In a system, energy itself is difficult to measure; instead, we measure changes in energy in units of joules (J):

$$\Delta E = E_{final} - E_{initial}$$

Internal energy is a *state function*, meaning that the quantity of E depends only on the present state of the system and not on how that state was achieved. A state function is independent of the path. For example, climbing a 24,000-foot mountain could be achieved two ways: the easy, long path or the short, difficult path. In the case of the mountain, the state function is altitude; regardless of your path, your altitude change will be the same: 24,000 feet. Internal energy can work the same way. A ΔE of 5 J could result from an increase in energy from 5 J to 10 J or an increase from 5 J to 15 J followed by a decrease in 5 J. ΔE is the same in both instances but achieved via different paths.

2. Provide another example of a state function, besides change in altitude.

The energy coming from chemical reactions is considered potential. Energy is absorbed when chemical bonds are broken, and it is released when bonds are formed. Depending on how much energy is absorbed and released during bond breaking and formation, the reaction may release or absorb energy. If overall energy is released from the system, it is known as an *exothermic process.* The exothermic process is where energy exits the system from the surroundings. *Endothermic processes* are the opposite; energy enters the system from the surroundings.

3. Define internal energy. Describe two ways in which internal energy may decrease.

For the rest of the chapter, we will focus on heat. *Heat* is the transfer of thermal energy. Heat is always the result of energy moving from a hot object to a cold object; this occurs until thermal equilibrium is established. *Thermal equilibrium* occurs when two objects reach the same temperature and heat is no longer transferred. When heated or cooled, objects change temperature, but different objects experience unique temperature changes for equal amounts of heat. This phenomenon is known as heat capacity. *Heat capacity* (C) is the amount of heat an object absorbs or releases to change its temperature by one unit; it is measured in J/K or J/°C. Specific heat capacity (aka specific heat) adds mass to the mix. *Specific heat capacity* (c) is the amount of heat 1 g of an object absorbs or releases to change its temperature by one unit, commonly measured in J/g·K or J/g·°C. The following equation employs specific heat (c) to calculate the heat (q) absorbed or released by a certain mass (m) for a change in temperature (ΔT):

$$q = mc\Delta T$$

4. Which would have a higher heat capacity, a mug of steaming water or a tub of water? Why?

5. Which would have a higher specific heat? Why?

Calorimetry is the study of heat transfer using a *calorimeter*, which is an isolated system that does not transmit matter or heat. Some calorimeters rely on water to absorb or release energy because scientists can easily determine the mass of water from volume and know the specific heat of water (4.184 J/g·K). To determine how calorimetry works, let's assume that we want to measure the amount of heat in a piece of hot iron. If we place the hot iron directly in water, we assume the amount of heat the metal loses $(-q_m)$ is equal to the amount of heat the water gains (q_w).

$$-q_m = q_w$$

Because $q = mc\Delta T$, we can substitute $-q_m$ and q_w with $mc\Delta T$.

$$-m_m c_m \Delta T_m = m_w c_w \Delta T_w$$

If the calorimeter absorbs a significant portion of the heat from the thermodynamic process occurring inside it, we must consider the heat capacity of the calorimeter. The heat absorbed by the calorimeter is the heat capacity of the calorimetry multiplied by the change in temperature:

$$q_{cal} = C_{cal} \Delta T$$

For our metal example, the water and calorimeter absorb its energy:

$$-q_m = q_w + q_{cal}$$

For chemical reactions, we can utilize the same principles as the metal example. Adding a chemical to the water in the calorimeter can cause a release or absorption of energy. If we were attempting to measure the heat of reaction (q_{rxn}), it would be equal to the amount of heat the solution (q_{soln}) lost or gained.

$$q_{soln} = -q_{rxn}$$

6. The heat of solution for KCl is 17.2 kJ/mol. What would be the amount of heat energy transferred if 1.0 mole of KCl were dissolved in H_2O?

7. A scientist used a calorimeter to determine the q_{rxn} for the reaction of H_2SO_4 and NaOH is –112 kJ. Would the amount of heat be the same, greater than, or less than if the student used a calorimeter with:

    (a) More insulation?

    (b) Less insulation?

A 'heating curve' is a graphical representation of the changes a substance undergoes as heat is added to it. These changes include temperature and phase changes, each of which has an associated enthalpy.

Review Questions

8. The flame of a candle and a campfire are the same temperature. Why do we use the campfire as a source of heat but not the candle?

9. Explain how heat capacity differs from specific heat.

10. Which requires more heat to heat from 25.0 °C to 50.0 °C, 30.0 g of H_2O (*l*) or 30.0 g of acetone (nail polish remover)? The specific heat of acetone is 2.15 J/g·K.

Application Questions

Use the chart of specific heats for some of the following questions.

Substance	Specific Heat (J/g·°C)
H_2O (l)	4.184
H_2O (s)	2.093
H_2O (g)	2.010
Diamond (C)	0.516
Fe	0.449
Cu	0.385
I_2	0.218
Pb	0.129

11. Calculate the change in internal energy for a system that does 42.0 J of work on the surroundings but absorbs 75.0 J of heat. Is the process endothermic or exothermic?

12. What is the specific heat of water? How much energy is required to raise 1.0 g of water by 1.0 K?

13. What is the energy required to raise 963 g of water by 1.0 K?

14. What is the heat capacity of 963 g of water? How does this relate to the energy required to raise 963 g of water by 1.0 K?

15. If 18.3 kJ of heat were added to 307 g of water, how much would the temperature increase?

16. Using the table of specific heats, which substance would heat the fastest if the same amount of heat were applied? Which would cool the slowest?

17. A 24.5 g piece of unknown substance requires 765 J to increase the temperature from 22.9 °C to 37.6 °C. What is the specific heat of this substance?

18. A 45.0 g piece of Cu at 85.0 °C is added to water at 25.0 °C; thermal equilibrium is established at 29.6 °C. What is the mass of the water?

19. A 21.0 g piece of Pb at 96.0 °C is added to 389 mL of water at 22.0 °C. At what temperature is thermal equilibrium established?

20. A mass of 2.98 g of NaOH is dissolved in 74.5 mL of H_2O to make a 1 M aqueous solution. The temperature increases by 10.6 °C. How much heat is absorbed or released by the reaction? Assume the solution has the same specific heat as liquid water.

21. In a calorimeter, 5.874 g of octane (C_8H_{18}) is combusted. The heat of combustion for octane is –5,508.9 kJ/mol. If the temperature of the calorimeter and its contents rises from 23.7 °C to 32.5 °C, what is the heat capacity of the calorimeter? What is the heat of combustion per gram of octane?

Figure Credit

CHAPTER 30

Energetics II

What is entropy?

What makes chemical reactions occur spontaneously?

How do we determine which reactions are spontaneous?

Learning Outcomes

- Calculate the enthalpy, entropy, and Gibbs free energy change of a process or reaction.
- Determine whether a process or reaction is spontaneous.
- Explain the relationship between enthalpy, entropy, and Gibbs free energy.
- Relate the equilibrium constant to Gibbs free energy.

Prerequisite Knowledge

- Chapter 25: Equilibrium Concepts
- Chapter 29: Energetics I

Background Information and Questions

Most chemical reactions are performed in open containers under atmospheric conditions at constant pressure. This allows for a simplification of calculations when determining the energy associated with chemical and physical processes. Measuring energy at constant pressure introduces a new thermodynamic term: *enthalpy*.

Enthalpy (H) is a state function like internal energy. It is considered the total heat within a system. The following equation illustrates the derivation of enthalpy from internal energy. Enthalpy varies from internal energy by excluding pressure–volume work.

As the derivation highlights, enthalpy is heat at constant pressure. In many cases, the enthalpy and heat are used interchangeably. The derivation below shows how internal energy can be used to derive enthalpy.

$$\Delta E = q_p + w_p$$

$$w_p = -P\Delta V$$

$$\Delta E = q_p - P\Delta V$$

$$q_p = \Delta E + P\Delta V$$

$$\Delta H = \Delta E + P\Delta V$$

$$\boldsymbol{\Delta H = q_p}$$

1. How is enthalpy related to heat?

Enthalpy is a common way that scientists report the heat energy associated with a chemical reaction. These are known as enthalpies of reaction (ΔH_r) and are typically reported in kilojoules (kJ) unless specified otherwise. Sometimes the "r" subscript is left off; a "c" subscript may accompany ΔH for enthalpy values referring to a combustion reaction, known as enthalpy of combustion (ΔH_c). Remember that the subscript just gives you a clue about what type of reaction the enthalpy is accompanying; for example, there are enthalpies of neutralization and enthalpies of solution, which refer to acid–base neutralization and dissolution of a compound, respectively.

2. Why do we measure the change in enthalpy and not just enthalpy?

Enthalpy can also be calculated for processes other than chemical reactions; enthalpy of fusion and enthalpy of vaporization refer to the heat energy associated with melting and vaporizing a substance. The following is an example chemical equation with the accompanying enthalpy value; this is known as a *thermochemical equation*, which is chemical reactions accompanied by their corresponding enthalpies of reaction.

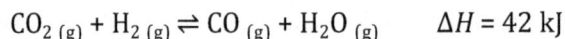

$$CO_{2\,(g)} + H_{2\,(g)} \rightleftharpoons CO_{(g)} + H_2O_{(g)} \qquad \Delta H = 42 \text{ kJ}$$

Enthalpies of formation (ΔH_f) are essentially "special" enthalpies of reaction for the formation of a compound reported in kJ/mol. Enthalpies of formation are calculated under standard state conditions, which is 1 bar, 1 M, and an unspecified temperature (most commonly 298 K). Standard state is denoted by a superscripted degree symbol. The following example is of the formation of nitrous oxide, along with its accompanying enthalpy of formation:

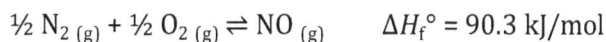

$$\tfrac{1}{2} N_{2\,(g)} + \tfrac{1}{2} O_{2\,(g)} \rightleftharpoons NO_{\,(g)} \qquad \Delta H_f^\circ = 90.3 \text{ kJ/mol}$$

3. Using the example equation above that illustrates the formation of NO (g), how much energy would be absorbed if 1.5 moles each of N_2 and O_2 were reacted?

Thermochemical chemical equations can be manipulated to determine new enthalpies. **Table 30.1** illustrates what can happen to the enthalpy when the equation is manipulated. One significant point about the thermochemical equation is that the enthalpy applies to the equation as it is written; therefore, 90.3 kJ is the enthalpy for the consumption of 0.5 mole of N_2. It also represents the enthalpy for the production of 1 mole of NO.

TABLE 30.1 Determination of Enthalpy Change by Manipulation of the Thermochemical Equation

Thermochemical Equation Manipulation	Resulting Enthalpy Change
Reverse the chemical equation	Change the sign of the enthalpy
Multiply/divide the chemical equation by a value	Multiply/divide the enthalpy by the same value

Enthalpies of reaction can be calculated from enthalpies of formation using the following equation. The idea behind the equation is that enthalpies of formation of the reactants must be overcome to break bonds and form products. Remember! Breaking bonds is endothermic, and the formation of bonds is exothermic. The overall reaction will be endothermic ($+\Delta H$) or exothermic ($-\Delta H$) based on whether more energy is used to break bonds or more energy is released when the new bonds are formed.

$$\Delta H_r^\circ = \sum n \Delta H_f^\circ (products) - \sum n \Delta H_f^\circ (reactants)$$

Enthalpies of reaction can also be calculated by applying *Hess's law*, which states that the sum of the ΔH_r° values of two or more reactions is the enthalpy of reaction for the overall process. Let's look at an example. We will produce CO_2 from graphite and oxygen. This reaction could occur through a two-step process with the addition of each oxygen.

The enthalpies of the first two reactions are displayed below. These two enthalpies will assist in the calculation of the enthalpy of reaction for the formation of carbon dioxide. The sum of those reactions is written underneath the underlined reaction. According to Hess's law, the sum of the enthalpies will be the enthalpy of the summed reaction; therefore, -110.5 kJ $+ -283$ kJ $= -393.5$ kJ. A graphical display of this process is presented in **Figure 30.1**.

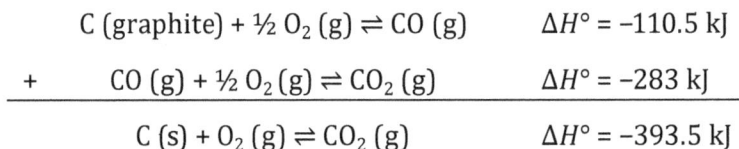

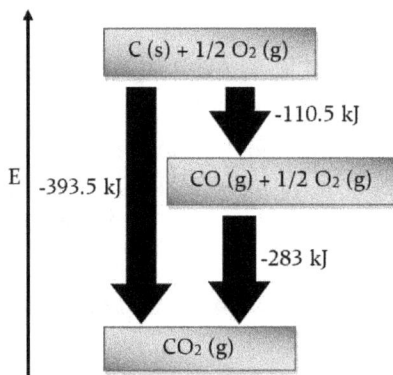

$$C \text{ (graphite)} + \tfrac{1}{2} O_2 \text{ (g)} \rightleftharpoons CO \text{ (g)} \qquad \Delta H° = -110.5 \text{ kJ}$$

$$+ \qquad CO \text{ (g)} + \tfrac{1}{2} O_2 \text{ (g)} \rightleftharpoons CO_2 \text{ (g)} \qquad \Delta H° = -283 \text{ kJ}$$

$$C \text{ (s)} + O_2 \text{ (g)} \rightleftharpoons CO_2 \text{ (g)} \qquad \Delta H° = -393.5 \text{ kJ}$$

FIGURE 30.1 Enthalpy of reaction for reaction of graphite and oxygen.

Why do these reactions happen? The reaction sequence in our example (Figure 30.1) releases -393.5 kJ of energy. Does the release of energy cause this reaction sequence to occur spontaneously, like an explosion? One would think that is a logical conclusion; however, some spontaneous reactions or processes absorb energy. For example, the dissolution of ammonium nitrate is spontaneous and has an enthalpy of solution of $+25.69$ kJ/mol. There must be another factor that contributes to spontaneity.

4. Using the reactions below, determine the enthalpy for $4 \text{ HF} + O_2 \rightarrow 2 \text{ F}_2 + 2 \text{ H}_2$.

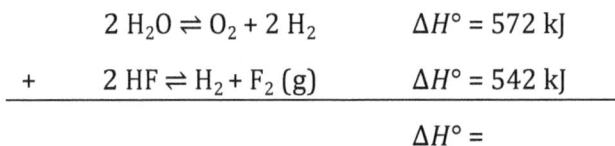

$$2 \text{ H}_2O \rightleftharpoons O_2 + 2 \text{ H}_2 \qquad \Delta H° = 572 \text{ kJ}$$

$$+ \qquad 2 \text{ HF} \rightleftharpoons H_2 + F_2 \text{ (g)} \qquad \Delta H° = 542 \text{ kJ}$$

$$\Delta H° =$$

Spontaneity, in terms of a chemical reaction, is when it occurs without outside intervention. Imagine that you drop a glass on the floor. It falls spontaneously; you don't have to push it. The potential energy is converted to kinetic energy as the glass falls; when it hits the floor, that kinetic energy is converted to other forms, like acoustic energy. The glass will also shatter and disperse itself. The energy transfer is very directional; the process cannot happen in reverse. We can conclude that spontaneity of a chemical reaction or process means that it must disperse energy, matter, or both. In other words, spontaneous processes always become more dispersed.

The *second law of thermodynamics* is built on this principle; it states that spontaneous processes increase the entropy of the universe. *Entropy (S)* is the thermodynamic term that is a measure of disorder. Entropy is also a state function.

Dispersal of energy results in an increase in molecular motion, which is thermal energy. This is inversely proportional to temperature. For a reversible process, like a phase change, entropy is represented by the following two equations. A reversible process is one that exists between two states and can be undone without changing the surroundings. It is at equilibrium at any point between the two states. One can also use the enthalpies of fusion and vaporization in place of q_{rev} because they are a measure of heat energy during a phase change at constant pressure.

$$S = \frac{q_{rev}}{T}$$

$$S_{fus/vap} = \frac{\Delta H_{fus/vap}}{T}$$

Another way to think about entropy is at the molecular level using *microstates (W)*. Think about microstates like numbered tiles that can be arranged. **Figure 30.2** illustrates six different microstates that the three tiles (numbered 1, 2, and 3) can be arranged to achieve. The more microstates (W), the higher the entropy of a system; the two variables are directly proportional. This is illustrated in the equation below, which features Boltzmann's constant (k).

$$S = k \ln W$$

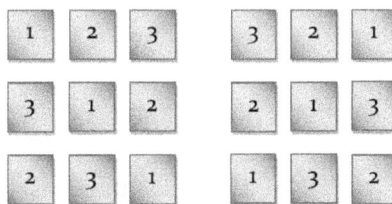

FIGURE 30.2 Microstates.

Entropy can be calculated for any process. Heating substances changes the enthalpy; phase changes also alter the enthalpy. Any substance on earth, under certain conditions, has an enthalpic value.

The *third law of thermodynamics* states that entropy is zero for a perfect crystal at zero Kelvin. Entropy for chemical reactions, typically nonreversible, can be derived much like enthalpy. There are tabulated entropy values for common chemicals at standard state conditions; these are used in the following equation. A positive ΔS results in an increase in enthalpy; a negative ΔS indicates a decrease.

$$\Delta S_r^\circ = \sum n \Delta S^\circ (products) - \sum n \Delta S^\circ (reactants)$$

Determining if a reaction is spontaneous means that one must determine if the ΔS(universe) increases:

$$\Delta S^\circ (universe) = \Delta S^\circ (system) + \Delta S^\circ (surroundings)$$

5. What must be dispersed for a process to be considered spontaneous?

A new thermodynamic quantity was developed to aid in the determination of spontaneity: *Gibbs free energy* (G). It includes both enthalpy and entropy. The following equation relates free energy changes to enthalpy and entropy changes of the system. It is derived from the ΔS(universe) equation above; that derivation is not shown.

$$\Delta G = \Delta H - T\Delta S$$

Gibbs free energy is also a state function. The change in Gibbs free energy of a reaction can be calculated from the tabulated values of Gibbs free energies of formation for reactants and products, just like enthalpy.

$$\Delta G_r^\circ = \sum n \Delta G_f^\circ (products) - \sum n \Delta G_f^\circ (reactants)$$

Table 30.2 indicates what values of ΔG_r° represent spontaneity.

TABLE 30.2 G$_r$ Values and Spontaneity

ΔG_r° Value	Spontaneity?
$\Delta G_r^\circ < 0$	Spontaneous
$\Delta G_r^\circ = 0$	At equilibrium
$\Delta G_r^\circ > 0$	Nonspontaneous

As you can see in Table 30.2, Gibbs free energy is related to the equilibrium of a chemical reaction. **Figure 30.3** graphically illustrates a spontaneous, product-favored reaction. Note the relationship between Q and ΔG. Also, note how the $\Delta G_r°$ is the difference between the $\Delta G_f°$ of the products and reactions.

6. For a reaction to be spontaneous, which must be larger, the Gibbs free energy of formation of products or the Gibbs free energy of formation of reactants?

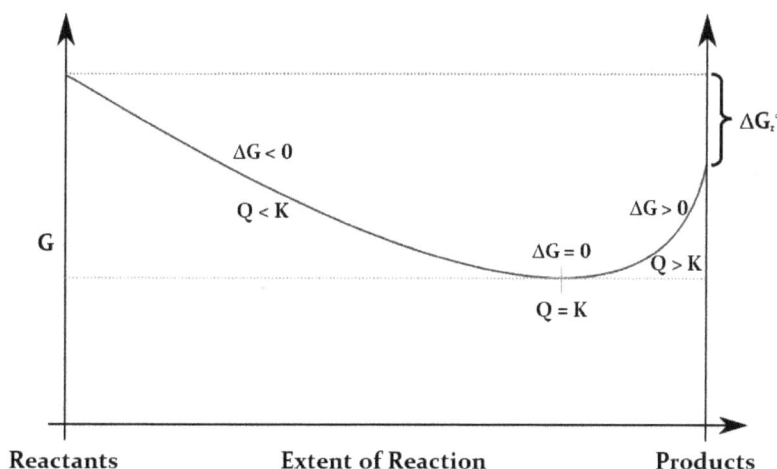

FIGURE 30.3 Relationship between Gibbs free energy and chemical equilibrium.

The reaction in Figure 30.3 will proceed from reactants to products spontaneously because the $\Delta G_f°$ of products is less than the $\Delta G_f°$ of reactants. It will proceed until it reaches equilibrium. At the moment $\Delta G = 0$, the reaction has reached equilibrium. If it proceeds past equilibrium, ΔG becomes positive, and making more products becomes nonspontaneous. The reverse reaction occurs spontaneously to reestablish equilibrium. This is all under standard state conditions. The following equation allows us to calculate the ΔG_r under nonstandard conditions, like a different temperature:

$$\Delta G_r = \Delta G_r° + RTlnQ$$

At equilibrium $\Delta G_r = 0$ and $Q = K$; therefore, the equation can be rearranged as follows, which allows for a direct calculation of K from $\Delta G_r°$ or vice versa:

$$⊠ = \Delta G_r° + RTlnK \qquad \Delta G_r° = -RTlnK$$

7. As a reaction approaches equilibrium, which value does ΔG approach?

8. Produce a graph for a reactant-favored, nonspontaneous reaction.

Review Questions

9. What is the driving force behind spontaneity in chemical reactions?

10. What is meant by the term "enthalpically driven"?

11. For a process at constant temperature and pressure, does the Gibbs free energy depend on the enthalpy, entropy, or both?

12. How is Gibbs free energy related to equilibrium?

Application Questions

Use the chart to answer the subsequent questions.

Substance	ΔH_f° (kJ/mol)	ΔS° (J/K·mol)	ΔG_f° (kJ/mol)
CO (g)	−110.5	197.9	−137.2
CO$_2$ (g)	−393.5	213.6	−394.4
Fe (s)	0	27.2	0
Fe$_2$O$_3$ (s)	−822.16	89.96	−740.98
H$_2$O (g)	−241.82	188.83	−228.57
NH$_3$ (g)	−46.19	192.5	−16.66
NO (g)	90.37	210.62	86.71
PbCO$_3$ (s)	−699.1	131.0	−625.5
PbO (s)	−217.3	68.70	−187.9

13. Write a balanced chemical equation for the formation of N$_2$O$_5$ from its elements in their standard oxidation states. Calculate the enthalpy change when 0.56 mole of N$_2$O$_5$ is formed.

14. Calculate the enthalpy of reaction for the oxidation of ammonia.

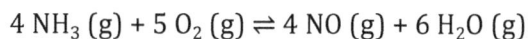

$$4\ NH_3\ (g) + 5\ O_2\ (g) \rightleftharpoons 4\ NO\ (g) + 6\ H_2O\ (g)$$

15. The synthesis of methanol from methane is:

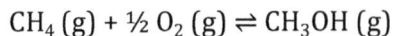

$$CH_4 \text{ (g)} + \tfrac{1}{2} O_2 \text{ (g)} \rightleftharpoons CH_3OH \text{ (g)}$$

Use the following reactions to determine the enthalpy of the reaction above.

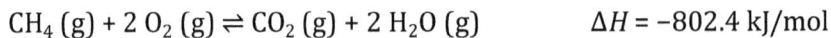

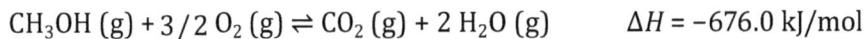

$$CH_4 \text{ (g)} + 2 O_2 \text{ (g)} \rightleftharpoons CO_2 \text{ (g)} + 2 H_2O \text{ (g)} \qquad \Delta H = -802.4 \text{ kJ/mol}$$
$$CH_3OH \text{ (g)} + 3/2 O_2 \text{ (g)} \rightleftharpoons CO_2 \text{ (g)} + 2 H_2O \text{ (g)} \qquad \Delta H = -676.0 \text{ kJ/mol}$$

16. Predict whether the following processes will result in an increase or decrease in entropy.

(a) melting snow

(b) condensation of water on a windshield

(c) sublimation of CO_2 (s)

(d) NH_4Br (s) $\rightleftharpoons$ NH_4Br (aq)

(e) $NaCl$ (s) $\rightleftharpoons$ $NaCl$ (g)

17. Predict the sign of ΔS for the following chemical reactions.

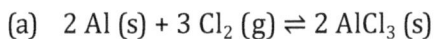

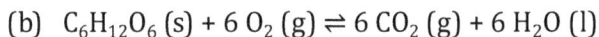

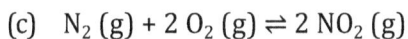

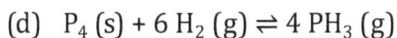

    (a) $2\ Al\ (s) + 3\ Cl_2\ (g) \rightleftharpoons 2\ AlCl_3\ (s)$

    (b) $C_6H_{12}O_6\ (s) + 6\ O_2\ (g) \rightleftharpoons 6\ CO_2\ (g) + 6\ H_2O\ (l)$

    (c) $N_2\ (g) + 2\ O_2\ (g) \rightleftharpoons 2\ NO_2\ (g)$

    (d) $P_4\ (s) + 6\ H_2\ (g) \rightleftharpoons 4\ PH_3\ (g)$

18. Calculate the entropy of reaction for the following chemical reaction.

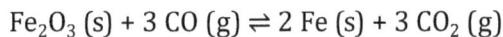

    $$Fe_2O_3\ (s) + 3\ CO\ (g) \rightleftharpoons 2\ Fe\ (s) + 3\ CO_2\ (g)$$

19. The table provided shows the possible sign combinations of ΔH and ΔS. Complete the table by entering +, –, or T in the ΔG column to correspond to the signs of ΔH and ΔS. Use T to indicate that the sign of ΔG depends on the temperature.

ΔH	ΔS	ΔG (+, –, or T)	Spontaneous? (Y, N, or depends on T)
+	+		
–	–		
+	–		
–	+		

20. Determine whether the following equation is spontaneous under standard state conditions.

$$PbCO_3 \text{ (s)} \rightleftharpoons PbO \text{ (s)} + CO_2 \text{ (g)}$$

21. A particular reaction has a ΔH of -80.0 kJ/mol and a ΔS of 104 J/mol·K. Would you expect the reaction to be spontaneous at 25 °C?

22. For the reaction $3 O_2 \text{ (g)} \rightleftharpoons 2 O_3 \text{ (g)}$, the ΔG_r° is 163 kJ/mol. Is the reaction product favored or reactant favored?

23. Calculate the values of K and ΔG for the following reaction.

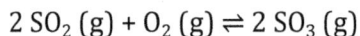

$$2 SO_2 \text{ (g)} + O_2 \text{ (g)} \rightleftharpoons 2 SO_3 \text{ (g)}$$

Figure Credit

CHAPTER 31

Introduction to Electrochemistry

How do batteries work?

What do the + and − on a battery indicate?

Learning Outcomes

- Determine the relative strength of oxidizing/reducing agents.
- Split redox reactions into their half-reactions.
- Interpret an electrochemical cell diagram: anode, cathode, salt bridge, electron flow.
- Write shorthand cell notation.
- Calculate cell potential.

Prerequisite Knowledge

- Chapter 12: Chemical Reactions I
- Chapter 13: Chemical Reactions II
- Chapter 15: Limiting Reactants and Molarity

Background Information and Questions

Electrochemistry is a branch of chemistry that is concerned with the relationship between electrical forces and chemical reactions. Recall from Chapter 13 that reactions where electrons are transferred from one species to another are oxidation–reduction reactions (redox). Redox reactions are at the heart of electrochemistry, and they can be split into half-reactions, one for reduction and one for oxidation. Consider the following redox reaction:

$$Zn_{(s)} + Cl_{2\ (g)} \rightarrow ZnCl_{2\ (s)}$$

Recall from Chapter 13 how to calculate oxidation numbers and determine which compounds are oxidized and which are reduced. In this case the zinc is oxidized (meaning it is the reducing agent—the charge on zinc goes from 0 to +2) and the chlorine is reduced (making it the oxidizing agent—the charge on chlorine goes from 0 to –1). Using that knowledge, the half-reactions are determined to be the following:

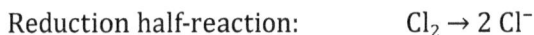

Reduction half-reaction: $Cl_2 \rightarrow 2\ Cl^-$

Oxidation half-reaction: $Zn \rightarrow Zn^{2+}$

However, these reactions are not balanced, so the next step is to add electrons to either side to balance out the charge:

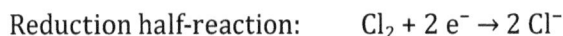

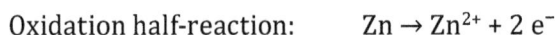

Reduction half-reaction: $Cl_2 + 2\ e^- \rightarrow 2\ Cl^-$

Oxidation half-reaction: $Zn \rightarrow Zn^{2+} + 2\ e^-$

When combined, these half-reactions once more yield the initial redox reaction, with the number of electrons on the reactant side and the number on the product side effectively "canceling out."

1. In the following reaction which compounds are reduced and which are oxidized?

$$3\ Mg + N_2 \rightarrow Mg_3N_2$$

(a) Which is the reducing agent and which is the oxidizing agent?

(b) Label each atom with its oxidation number for this reaction.

(c) Write (and label!) the oxidation and reduction half-reactions.

The movement of electrons, which are charged particles, in a redox reaction has an associated electric potential (E). The potentials for each half-reaction can be estimated using a standard hydrogen electrode (SHE) and the values placed into a table for ease of use (see **Table 31.1**). Typically, reduction potentials are listed, while oxidation potentials are calculated from the reduction potentials by flipping the sign (positive or negative).

TABLE 31.1 Standard Reduction Potentials

Reaction	Reduction Potential
$F_{2\,(g)} + 2\,e^- \rightleftharpoons 2\,F^-_{\,(aq)}$	2.87
$Au^+ + e^- \rightleftharpoons Au_{\,(s)}$	1.83
$Ce^{4+}_{\,(aq)} + e^- \rightleftharpoons Ce^{3+}_{\,(aq)}$	1.44
$Cl_{2\,(g)} + 2\,e^- \rightleftharpoons 2\,Cl^-_{\,(aq)}$	1.40
$2\,Hg^{2+}_{\,(aq)} + 2\,e^- \rightleftharpoons Hg_2^{2+}_{\,(aq)}$	0.911
$Ag^+_{\,(aq)} + e^- \rightleftharpoons Ag_{\,(s)}$	0.800
$Cu^+_{\,(aq)} + e^- \rightleftharpoons Cu_{\,(s)}$	0.52
$Cu^{2+}_{\,(aq)} + 2\,e^- \rightleftharpoons Cu_{\,(s)}$	0.342
$Sn^{4+}_{\,(aq)} + 2\,e^- \rightleftharpoons Sn^{2+}_{\,(aq)}$	0.154
$2\,H^+_{\,(aq)} + 2\,e^- \rightleftharpoons H_{2\,(g)}$	0
$Fe^{3+}_{\,(aq)} + 3\,e^- \rightleftharpoons Fe_{\,(s)}$	−0.037
$Sn^{2+}_{\,(aq)} + 2\,e^- \rightleftharpoons Sn_{\,(s)}$	−0.14
$Ni^{2+}_{\,(aq)} + 2\,e^- \rightleftharpoons Ni_{\,(s)}$	−0.257
$Fe^{2+}_{\,(aq)} + 2\,e^- \rightleftharpoons Fe_{\,(s)}$	−0.41
$Zn^{2+}_{\,(aq)} + 2\,e^- \rightleftharpoons Zn_{\,(s)}$	−0.762
$Cr^{2+} + 2\,e^- \rightleftharpoons Cr_{\,(s)}$	−0.90
$Al^{3+}_{\,(aq)} + 3\,e^- \rightleftharpoons Al_{\,(s)}$	−1.68
$Ba^{2+} + 2\,e^- \rightleftharpoons Ba_{\,(s)}$	−2.92
$Li^+_{\,(aq)} + e^- \rightleftharpoons Li_{\,(s)}$	−3.04

A reactant in a reaction with a positive potential is a good oxidizing agent and is easy to reduce (as you would expect, it is easy to reduce F_2 to F^-). Another way to think of this is that the more positive the potential for a particular species, the greater the species' affinity for electrons. Similarly, a product of a reaction with a negative potential is a good reducing agent and easy to oxidize (it is easy to oxidize Li to Li^+).

   2. Which is a better oxidizing agent according to Table 31.1, Au^+ or Fe^{3+}?

The "chemical energy" that results from a spontaneous redox reaction can be turned into electrical energy with the use of an electrochemical cell called a galvanic cell (aka voltaic cell). Electrodes (aka plates and terminals) are either of two substances that enable an electric current to flow in the presence of an electrolyte. The cathode is the electrode at which reduction takes place (*Tip: C and R are both consonants*), and the anode is the electrode at which oxidation takes place (*Tip: A and O are both vowels*). The cell consists of these two electrodes, each of which is in contact with an electrolyte solution, and they are often connected by a salt bridge. The cathode is the positive (+) terminal, and the anode is the negative (–) terminal. An electrochemical cell is shown in **Figure 31.1**.

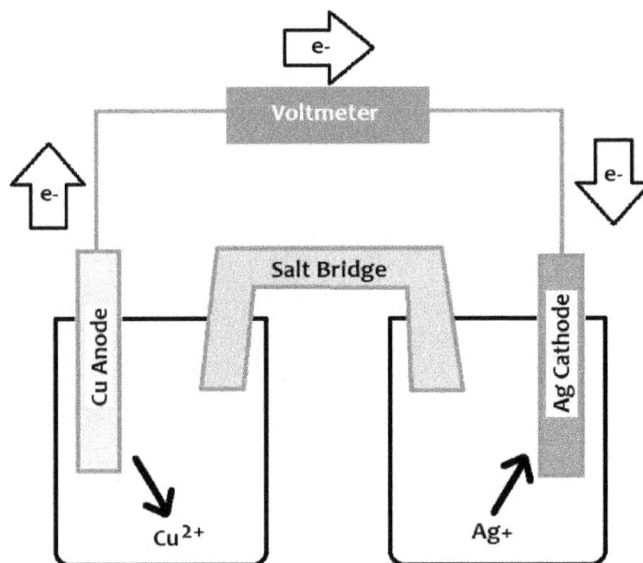

FIGURE 31.1 Electrochemical cell.

What is happening in the galvanic cell in Figure 31.1? First, let's look at the reactions taking place at the anode. At the anode we have $Cu \rightarrow Cu^{2+} + 2e^-$, where one of the copper atoms from the anode loses its electrons (which flow up the wire) and releases a copper cation (Cu^{2+}) into the solution. These electrons create a current as they move across the wire to the solid silver cathode, where they will interact with Ag^+ ions in solution and form more solid silver (which is deposited on the cathode). Therefore, the reaction occurring at the cathode is $Ag^+ + e^- \rightarrow Ag$. Not shown are the reactions of the salt bridge with the solutions in the beakers—when the Cu^{2+} enters the solution from the anode, a cation from the salt bridge (such as Cl^-) will balance the charge in the solution. A similar reaction occurs with the salt bridge when the Ag^+ is deposited on the cathode.

The purpose of the salt bridge is to prevent charge buildup. It is not shown here, but if the salt bridge contained KNO_3 then NO_3^- ions would shift into the copper solution (to balance out the added Cu^{2+} charges) while K^+ ions moved into the silver solution (to balance out the Ag^+ ions that are leaving solution as Ag).

So how much electrical potential is present in this cell? It can be measured by a voltmeter, but we can calculate it using the reduction potentials listed in Table 31.1.

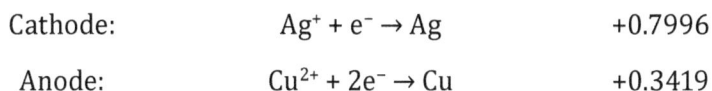

| Cathode: | $Ag^+ + e^- \rightarrow Ag$ | +0.7996 |
| Anode: | $Cu^{2+} + 2e^- \rightarrow Cu$ | +0.3419 |

To calculate the standard reduction potential for the entire reaction, we will use these values with the following equation:

$$E^0_{cell} = E^0_{cathode} - E^0_{anode}$$

$$E^0_{cell} = +0.7996 - (+0.3419) = +0.4577$$

> Note that oxidation, not reduction, is actually occurring at the anode. Some textbooks will teach the equation cathode PLUS anode and require the student to "flip" the sign at the anode, in this case from +0.34 to −0.34, to get the correct value and then add the two values together (the result is the same, just arrived at by different logic!).

Galvanic cells rely on spontaneous reactions to produce electricity, which means that the result for E^0_{cel} MUST be positive or else electrons will not flow. If E^0_{cel} is negative, then the reaction will not proceed naturally, and electrons will not move.

3. Does oxidation occur at the cathode or the anode?

4. Does reduction occur at the cathode or the anode?

5. Which direction do electrons move, from the anode to the cathode or from the cathode to the anode?

6. A galvanic cell is created using the following two half-reactions. What is the cell potential (E^0_{cel})?

   Hint: Look up the values on the table, then say cathode-anode!

$$Zn^{2+} + 2\ e^- \rightleftharpoons Zn$$

$$Cu^{2+} + 2\ e^- \rightleftharpoons Cu$$

The final concept to consider in this chapter is how to write out the information in a galvanic cell in shorthand. Describing or drawing a cell would be time consuming, so a *cell diagram* has been developed to provide the relevant information. A single line represents the phase boundary at the electrode, and the double line in the middle represents the salt bridge.

Anode | Anode ion solution || Cathode ion solution | Cathode

For the cell shown in Figure 31.1, the shorthand would be as follows:

Cu (s) | Cu^{2+} (aq) || Ag$^+$ (aq) | Ag (s)

If concentrations of the solutions are known, then they will be listed in the shorthand notation as well. This will come into play in the next chapter, so stay tuned!

7. Write the shorthand notation for the galvanic cell described in question 6.

Review Questions

8. Given what you now know about converting chemical energy into electrical energy, how would a battery work conceptually?

9. Would a battery be a galvanic cell? Why or why not?

Application Questions

Use the following reaction to answer the following questions:

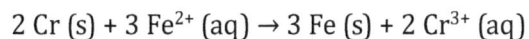

$$2\ Cr\ (s) + 3\ Fe^{2+}\ (aq) \rightarrow 3\ Fe\ (s) + 2\ Cr^{3+}\ (aq)$$

10. Write out the two half-reactions for the reaction.

11. Which half-reaction would occur at the anode and which at the cathode?

12. What is the standard cell potential for the galvanic cell based on this reaction?

13. Write out the shorthand notation for the galvanic cell based on the reaction.

CHAPTER 32

Advanced Electrochemistry

What happens if you reverse the cathode and anode?

How does electrochemistry relate to other areas of chemistry?

How does chrome plating work?

Learning Outcomes

- Apply the Nernst equation to solve advanced electrochemistry problems.
- Describe the relationship between cell potential, free energy, and the equilibrium constant.
- Explain the difference between an electrolytic and a voltaic cell.

Prerequisite Knowledge

- Chapter 25: Equilibrium Concepts
- Chapter 30: Energetics II
- Chapter 31: Introduction to Electrochemistry

Background Information and Questions

In our introduction to electrochemistry in Chapter 31, we assumed standard conditions, which means that all the galvanic cells had concentrations of 1 M. In reality this is rarely the case. To determine the cell potential for a more complex galvanic cell, we need to use the Nernst equation:

$$E = E^0 - \frac{0.0592}{n} \log Q$$

This equation is derived from the equation for Gibbs free energy discussed in Chapter 30. In this equation, E is the cell potential, $E°$ is the standard cell potential (which you learned to calculate in the previous lesson using cathode minus anode), n is the number of electrons that are moving in the reaction, and Q is the reaction quotient, products over reactants, from Chapter 25 on equilibrium. Whew, that's a lot! We won't go through the entire derivation here, but the end result is that galvanic cells must be spontaneous reactions. Spontaneous reactions have a negative value for ΔG and a positive value for E.

There is one way to simplify the equation, which occurs only at equilibrium. At equilibrium $Q = K$ and E = 0. Therefore, the Nernst equation simplifies to the following form:

$$E^0 = \frac{0.0592}{n} \log K$$

1. Why do you think E = 0 at equilibrium?

2. Why do galvanic cells need to be spontaneous?

Here's one example of how to use the Nernst equation. Consider the following reaction:

$$Co\ (s) + Fe^{2+}(aq)\ (1.5\ M) \rightarrow Co^{2+}(aq)\ (0.5\ M) + Fe\ (s)$$

Is this reaction spontaneous? Will it result in a galvanic cell? We can answer this by finding the value for E using the Nernst equation:

$$E = E^0 - \frac{0.0592}{n} \log Q$$

First, let's determine $E°$. We can do this by calculating cathode minus anode, as in the previous chapter:

$Co^{2+} + 2\ e^- \rightarrow Co\ (s)$	$E° = -0.28$
$Fe^{2+} + 2\ e^- \rightarrow Fe\ (s)$	$E° = -0.447$

Looking at the original reaction, we can see that the cathode is Fe and the anode is Co; therefore:

$$cathode - anode = -0.447 - (-0.28) = -0.17 = E°$$

Next, we can determine a value for n, the number of moles of electrons that are moving in this reaction. Looking at the original reaction, we can see that when cobalt loses two electrons, iron gains them, so overall there are two electrons moving for each reaction that takes place. Therefore, $n = 2$.

Finally, we must calculate Q using products over reactants. Our product in this case is the cobalt ion, which has a concentration of 0.5 M. Our reactant is the iron ion at 1.5 M. Note that the solids are not part of the equilibrium. Therefore:

$$Q = \frac{[Products]}{[Reactants]} = \frac{0.5}{1.5} = 0.33$$

Now we have all of the required information to calculate E using the Nernst equation!

$$E = -0.17 - \frac{0.0592}{2} \log(0.33) = -0.15$$

Because E has a negative value, we can say that this cell will NOT be spontaneous and would not serve as a galvanic cell!

3. Would the following result in a galvanic cell? *Hint: Just follow the steps above! If you need a refresher on how to find* E°, *this problem was used as an example in the previous lesson.*

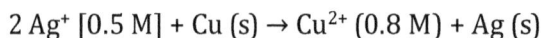

$$2\ Ag^+\ [0.5\ M] + Cu\ (s) \rightarrow Cu^{2+}\ (0.8\ M) + Ag\ (s)$$

What the Nernst equation illustrates is that the potential of an electrochemical cell, Gibbs free energy, and equilibrium are all related both conceptually and mathematically.

$$\Delta G^0 = -RT\ln K \qquad \text{relates G to K}$$

$$\Delta G^0 = -nFE^0_{cell} \qquad \text{relates G to E}$$

$$E^0_{cell} = \left(\frac{RT}{nF}\right)\ln K \qquad \text{relates E to K}$$

Note: F is the Faraday constant (96,485 C/mol), and R is the ideal gas constant (8.31415 J/mol·K)—always check your units to make sure you are using the correct values!

4. Conceptually, how do G, K, and E relate to each other? When a system is at equilibrium, what does that indicate about the free energy? If the free energy is negative, what does that indicate about the electrochemical potential? And so on.

If an electrochemical cell does not result in a galvanic cell, what *does* it result in? An *electrolytic cell* is the answer. Conceptually, you may think of an electrolytic cell as the inverse of a galvanic cell—instead of using a chemical reaction to generate electricity, an electrolytic cell uses electricity to drive a chemical reaction. Electrolytic cells have a negative value for E and are not spontaneous.

Common uses of electrolytic cells are metal plating and purification of metals. Recall that in a galvanic cell, a solid metal is deposited on the cathode. In an electrolytic cell, current can be used to force a specific metal (such as chrome or silver) to deposit and coat an item such as a spoon with the desired metal.

5. In all electrochemical cells, reduction occurs at the cathode and oxidation at the anode. In a galvanic cell, the anode is negative and the cathode is positive. Will that be the same for an electrolytic cell? Explain your reasoning for why or why not.

Review Questions

6. A galvanic cell in which the same reaction occurs at both the cathode and the anode (but in reverse) is called a concentration cell. In these cells the only difference at cathode and anode is the concentration of ion.

   (a) What would $E°$ be for a concentration cell? Why?

(b)   Would it be possible to use an electrolytic cell instead of a galvanic cell to make a battery? Why or why not?

7. Would the following reactions be spontaneous or nonspontaneous? Circle!

   +E $-\Delta G$ $Q < K$ spontaneous OR nonspontaneous?

   $-E$ $+\Delta G$ $Q > K$ spontaneous OR nonspontaneous?

8. Why is 2 H^+ (aq) + 2 e^- ⇌ H_2 (g) listed as having a reduction potential of 0 on most tables that list it? *Hint: You may need to combine your new knowledge from this lesson with facts about SHE from the previous chapter to answer this one!*

Application Questions

Use the following reaction to answer the following questions:

$$Pb^{2+} + Zn\ (s) \rightarrow Zn^{2+} + Pb\ (s)$$

9. If the concentration of Pb^{2+} is 0.00020 M, and the concentration of Zn^{2+} is 3.0 M, what is the cell potential for this reaction? *Hint: Find the standard cell potential first, then use the Nernst equation!*

10. What is ΔG for the reaction? Hint: The Faraday constant is 96485 C/mol e⁻.

11. What is K for the reaction?

CHAPTER 33

Nuclear Chemistry

What is radiation?

How does fusion differ from fission?

What are daughter products?

Learning Outcomes

▪ Rationalize nuclear stability using n:p ratios and nuclear binding energy.
▪ Recognize the types of nuclear radiation.
▪ Explain why a nuclide undergoes a specific type of radioactive decay.
▪ Write nuclear reactions based on a specific type of radioactive decay.

Prerequisite Knowledge

▪ Chapter 3: Atomic Structure
▪ Chapter 4: Electronic Structure I

Background Information and Questions

Nuclear bombs are one of the first things that many people associate with the discipline of chemistry. Nuclear chemistry has been utilized for more than death and destruction; in fact, it is used in many life-saving health care techniques today. Regardless of the application, nuclear chemistry starts with the nucleus.

In this lesson you may see terminology like *nucleon*, which is a proton or a neutron, or *nuclides*, which refers to a single type of nucleus. Nucleons and nuclides are the centers of focus for this lesson. A nuclide is positively charged due to the protons, so what holds it together? A strong attraction is necessary to overcome the repulsions of the positive protons; this is known as the *strong nuclear force*. We measure this force with *nuclear binding energy* (E_b), which is the energy required to disassemble a nucleus.

Recall that protons and neutrons are the subatomic particles that reside in the nucleus, and different numbers of neutrons result in isotopes. Shown below is lithium-7, a lithium isotope with a 92% abundance, which has three protons, four neutrons, and three electrons. This isotope has an experimentally determined mass of 7.0160 amu. If we calculate the mass of the isotope using the masses of protons, neutrons, and electrons, we get a calculated isotopic mass of 7.0584 amu:

$$^{7}_{3}Li$$

Isotopic mass = 7.0160 amu

(3 × 1.0073 amu) + (4 × 1.0087 amu) + (3 × 0.00055 amu) = 7.0584 amu

protons   neutrons   electrons

The difference in the masses (7.0584 – 7.0160 = 0.0424 amu) is known as the *mass defect*. The calculated mass is larger than the experimentally determined one because some of the calculated mass is converted to energy as the atom forms. The "missing" mass is equated to the nuclear binding energy. The mass–energy relationship is explained by Einstein's equation:

$$E = mc^2$$

where E is energy, m is mass, and c is the speed of light in a vacuum.

1. What is the term for the force that holds nuclei together? Why must it be strong?

2. Why is matter not conserved in a nuclear reaction?

Of all the isotopes that exist, only certain ones are stable; unstable nuclei are *radioactive*. One way of determining nuclear stability is to consider the ratio of neutrons to protons. As shown in **Figure 33.1**, nuclei with small numbers of protons (< 20) are stable at a 1:1 neutron:proton (n:p) ratio. However, as the number of protons increases, more stable nuclei prefer a higher n:p ratio; in other words, there are more neutrons than protons.

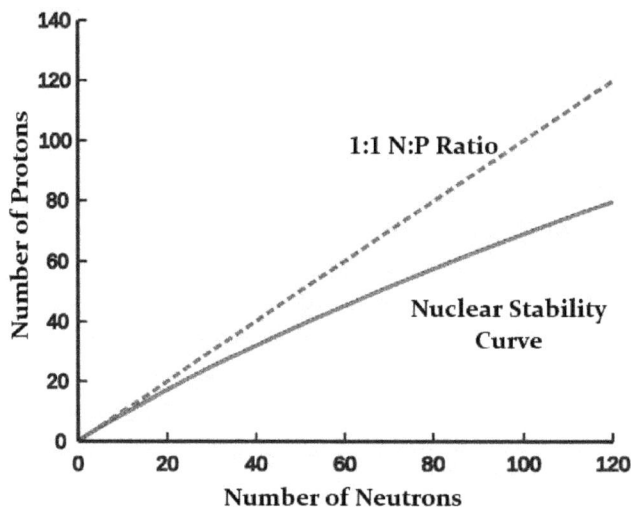

FIGURE 33.1 Nuclear stability.

If we plot the number of nucleons versus the E_b per nucleon (see **Figure 33.2**), we notice that stability increases as the nucleus gets larger, but the largest nuclei are not quite as stable, either. Isotopes beyond lead-208, where many decay series end, or bismuth-209 are all unstable and radioactive. The most stable nucleus is iron-56, which is marked by the vertical line. The plot in Figure 33.2 also illustrates that larger nuclei prefer fission (splitting a nucleus), while smaller nuclei are more likely to undergo fusion (uniting two nuclei).

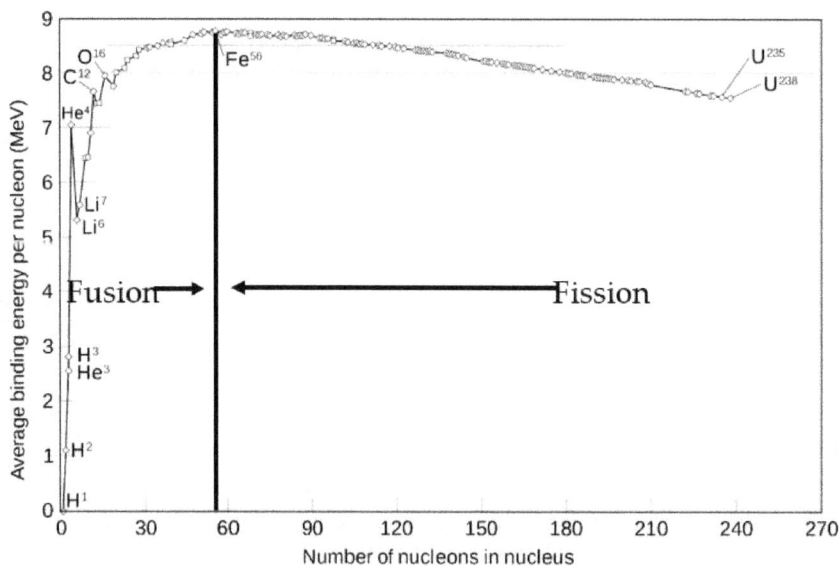

FIGURE 33.2 Number of nucleons versus the E_b per nucleon.

3. Why is iron-56 one of the more stable nuclides?

The stability of a nuclide and its neutron:proton ratio determine what type of radiation, if any, a particular nuclide may emit. **Table 33.1** illustrates the types of radiation. An alpha particle is essentially a helium nucleus, a beta particle is an electron, and a positron is a positive version of a beta particle.

TABLE 33.1 Types of Radiation

Name	Symbol	Representation
Alpha particle	^4_2He	
Beta particle	$^0_{-1}\text{e}$	
Positron	$^0_{+1}\text{e}$	
Neutron	^1_0n	
Proton	^1_1p	
Gamma rays	γ	

Nuclear decay can also emit protons, neutrons, and gamma rays. Large elements ($Z \geq 83$) tend to utilize alpha particles as a way to lower their atomic numbers (Z) and get closer to the band of stability around $Z = 56$. A high n:p ratio isotope (above the band of stability) can undergo beta emission; this leads to a lower n:p ratio because the Z increases by 1 and the mass number (A) remains constant. A low n:p ratio isotope (below the band of stability) prefers to decay by positron emission or electron capture. This type of emission leads to a lower Z but a constant A value.

Nuclear reactions are the way that we represent the conservation of particle in a nuclear process. They must be balanced just like chemical reactions. Here are some rules for balancing nuclear equations.

I. The sum of the mass numbers of reactants must equal the sum of the mass numbers of products.

II. The sum of the atomic numbers of reactants must equal the sum of the atomic numbers of products.

III. The sum of the charges of reactants must equal the sum of the charges of products.

Let's observe a nuclear reaction resulting in alpha emission. Note how this reaction follows the three rules above.

$$^{230}_{90}\text{Th} \rightarrow {}^{226}_{88}\text{Ra} + {}^{4}_{2}\text{He}$$

Table 33.2 lists some common nuclear reactions; it also illustrates the changes in A and Z for the original nuclide on the reactant side of the equation.

TABLE 33.2 Common Nuclear Reactions Types

Type of Decay	Nuclear Reaction Example	Change in A or Z
Alpha decay	$^{230}_{90}\text{Th} \rightarrow {}^{226}_{88}\text{Ra} + {}^{4}_{2}\text{He}$	A decreases by 4, Z decreases by 2.
Beta decay	$^{214}_{82}\text{Pb} \rightarrow {}^{214}_{83}\text{Bi} + {}^{0}_{-1}\text{e}$	A is unchanged, Z increases by 1.
Positron emission	$^{207}_{84}\text{Po} \rightarrow {}^{207}_{83}\text{Ra} + {}^{0}_{+1}\text{e}$	A is unchanged, Z decreases by 1.
Electron capture	$^{81}_{37}\text{Rb} + {}^{0}_{-1}\text{e} \rightarrow {}^{81}_{36}\text{Kr}$	A is unchanged, Z decreases by 1.

4. Label the type of decay, and balance the nuclear reactions below:

(a) $^{3}_{1}\text{H} \rightarrow$ _____ $+ {}^{0}_{-1}\text{e}$

(b) $^{222}_{86}\text{Rn} \rightarrow$ _____ $+ {}^{4}_{2}\text{He}$

We can measure the amount of decay through half-life, just like we did in Chapter 23. Half-life is the amount of time it takes for half of the isotope to decay. For nuclear reactions, this will always be first order. More stable isotopes tend to have longer half-lives than unstable ones. Uranium-238 has a half-life of 4.5 billion years, whereas polonium-214 has a half-life of 164 microseconds. Isotopes used in medical applications—like iodine-131, technetium-99, sodium-24, and thallium-201—tend to half shorter half-lives to minimize the amount of radiation exposure for the patient and people interacting with the patient.

Unstable isotopes decay to form more stable isotopes. Certain isotopes decay through a series of alpha and beta emissions called a *radioactive decay series*. The decay of uranium-238 is illustrated in **Figure 33.3**. Decay products are sometimes called *daughter products.* Note how certain isotopes undergo alpha emission until a certain n:p ratio is achieved, then the subsequent emission is a beta particle. Also, note the half-lives of the isotopes.

5. How do we use half-life with respect to nuclear chemistry?

Nuclear transmutation is the conversion of one nuclide to another nuclide; this process can occur by radioactive decay or by the reaction of a nucleus with another particle. Two types of reactions that result in transmutation are nuclear fusion and nuclear fission.

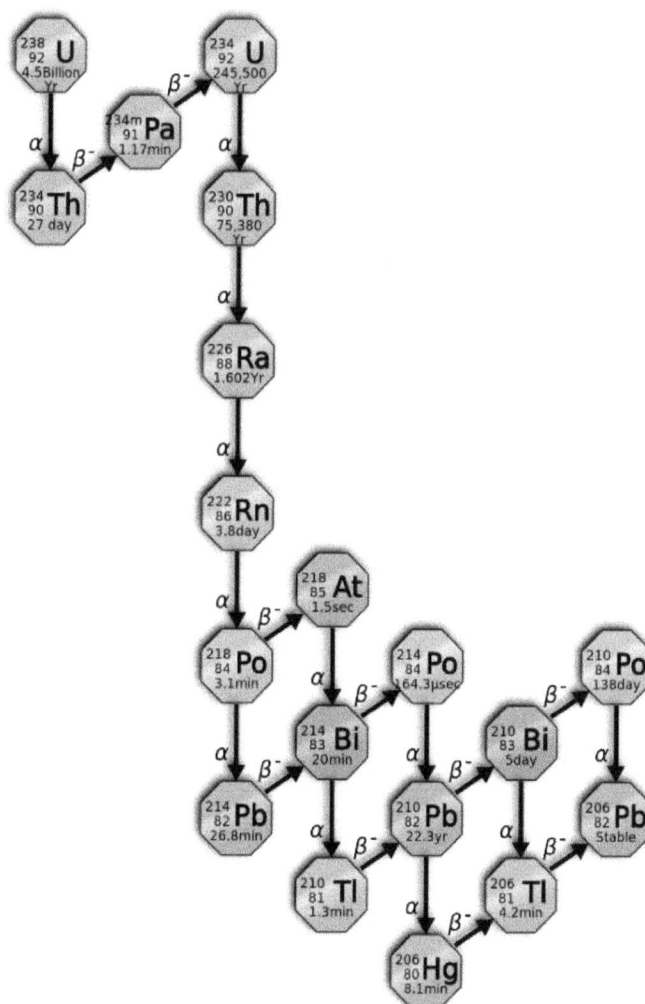

FIGURE 33.3 Radioactive decay series of uranium-238.

Nuclear fission is the splitting of a nuclide. **Figure 33.4** is an illustration of the splitting of a uranium-235 nucleus. This reaction was used in the atomic bomb dropped on Hiroshima and sets off a chain reaction. It absorbs a neutron, which reduces its stability. The uranium-236 nucleus undergoes fission to produce barium-144, krypton-89, and three neutrons. The three neutrons are absorbed by more uranium-235 nuclei, and more splitting reactions occur. This chain reaction releases a huge amount of energy. Controlled nuclear fission is also used in nuclear reactors to generate energy. Nuclear fusion also releases energy. Figure 33.4 also illustrates the fusion of deuterium (^{2}H) and tritium (^{3}H) to form helium. Nuclear fusion is how our sun generates its energy; it fuses hydrogen nuclei to form helium nuclei in a process similar to the illustration in the figure.

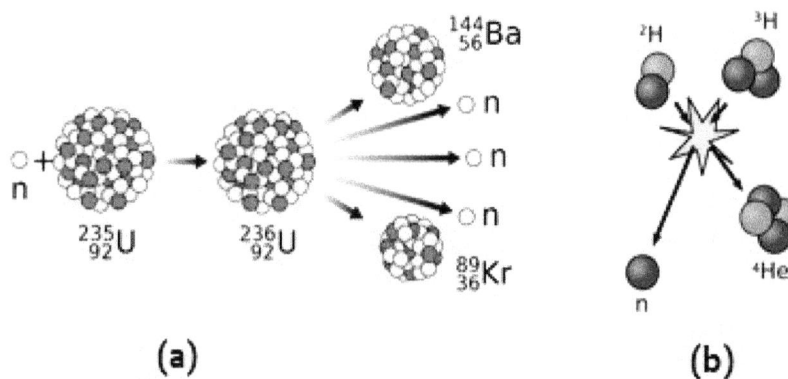

FIGURE 33.4 Splitting of a uranium-235 nucleus (a) and fusion of deuterium and tritium (b).

6. Compare and contrast nuclear fission and nuclear fusion.

Review Questions

7. What are the two parameters that are utilized to determine the stability of a nuclide?

8. Why do nuclei under spontaneous radioactive decay?

9. What is a radioactive decay series, like the one for U-238 in Figure 33.3?

Application Questions

10. What are the number of protons and neutrons in the following nuclei?

    (a) $^{239}_{94}Pu$

    (b) $^{40}_{19}K$

    (c) $^{4}_{2}He$

11. Using $E = mc^2$, calculate the nuclear binding energy in J/mol for carbon-13, which has an isotopic mass of 13.003 amu.

12. Complete the following nuclear reactions AND identify the type of radiation that is involved:

    (a) $^{81}_{37}Rb + ^{0}_{-1}e \rightarrow$?

(b) $^{40}_{20}\text{Ca} + ^4_2\text{He} \rightarrow 2\ ^1_0\text{n} + ?$

(c) $^{222}_{86}\text{Rn} \rightarrow ^{218}_{84}\text{Po} + ?$

13. Which of the following nuclides decay by $^0_{+1}\text{e}$ emission? Explain why.

(a) $^{238}_{92}\text{U}$

(b) $^{13}_6\text{C}$

(c) $^{230}_{90}\text{Th}$

(d) $^{15}_7\text{N}$

14. Plutonium-241 undergoes radioactive decay beginning with $^{241}_{94}\text{Pu}$ through the following sequence: β^-, α, α, β^-, α, α, β^-, α, α, α, α, β^-, β^-, and α. Draw a radioactive decay series and label each daughter product in the decay.

Figure Credits